Cracking the Technician License Exam

Copyright © 2019 by Critical Communications Media LLC

All rights reserved. No part of this publication may be reproduced, distributed, or transmitted in any form or by any means, including photocopying, recording, or other electronic or mechanical methods, without the prior written permission of the publisher, except in the case of brief quotations embodied in critical reviews and certain other noncommercial uses permitted by copyright law. For permission requests, write to the publisher, addressed "Attention: Permissions Coordinator," at the address below.

8605 Santa Monica Boulevard #92424
West Hollywood, CA 90069
Or visit: www.hamradioprep.com

Ordering Information:
Special discounts are available on quantity purchases by corporations, associations, and others. For details, contact the publisher at the address below.

Telephone: (833) HAMEXAM or (833) 426-3926
Email: contact@hamradioprep.com
Printed in the United States of America
ISBN: 9781797068596

This book is dedicated to our biggest fans a.k.a our parents: Peggy, Sherrie, Eric, and Jim.

We owe so much to you and could never express how much you mean to us.

ACCESS TO ONLINE COURSE ($25 VALUE!)

With the purchase of this book, you also have access to Ham Radio Prep's online course. With the online course you'll gain access to:

- Videos that teach you concepts that a book can't animate
- *Unlimited* Practice Final Exams
- Ability to study on the go with your desktop, mobile, and tablet
- Access to the *quiz generator* for another way to practice

To gain access to *Ham Radio Prep's Online Technician Course* simply follow the steps below:

1) Go to the back of this book and redeem the unique coupon provided that will give you **100% off the course**
2) Go to www.HamRadioPrep.com and add the online Technician Course to your shopping cart
3) Create an account with your name and email address
4) Use the coupon code at checkout (NO credit card required!)

BEFORE YOU GET STARTED

Thank you for choosing Ham Radio Prep to study for your Technician Exam. A couple of things you should know before you start studying with us:

1) Throughout the lessons, you will see words that have been **bolded**. Make sure to focus on these words as these are the correct answers on the actual FCC exam. We found that our students pass at much higher rates when they are able to recognize the correct answer in addition to learning the material.

2) Make sure that you're scoring at least an **80%** on both your end of chapter quizzes, as well as, your final exam before you register for test day!

3) We love to brag about our students! When (not if) you pass the exam email us a picture of you holding up your **Completion of Examination (*CSCE*)** and we will send you a free gift on us! You might even show up on our Facebook or Twitter page :)

4) We are **HERE FOR YOU!** When we started our journey in Ham Radio we were disappointed by the endless free apps, study guides, and online courses that were seemingly made in 1995. It was and still is, our mission to create the BEST way to study for the Technician Exam. If you have any questions, comments, or concerns about your purchase, please don't hesitate to reach out to us by:

Call or Texting: (833) HAMEXAM
Emailing: contact@hamradioprep.com
Writing to Us:
Critical Communications Media LLC
8605 Santa Monica Blvd #92424
West Hollywood, CA 90069

Table of Contents

Chapter 1: Introduction To Amateur Radio 9
Chapter 2: Station Operation 28
Chapter 3: Radio Wave Characteristics 41
Chapter 4: Amateur Radio Practices 53
Chapter 5: Electrical Principles 63
Chapter 6: Electrical Components 78
Chapter 7: Station Equipment 88
Chapter 8: Modulation Modes 99
Chapter 9. Antennas and Feedlines 113
Chapter 10. Electrical Safety 121

Final Exam 1 130
Final Exam 2 140
Final Exam 3 151
Chapter Quiz Answers 161
Answer Key for Final Exams 163

Chapter 1: Introduction To Amateur Radio

Amateur Radio Service purpose

Before jumping into the details of Ham radio, let's talk about why the Amateur Radio service exists in the first place. According to the FCC, the purpose of the Amateur Radio Service is **advancing skills in the technical and communication phases of the radio art.**

When you're taking the test, the key word here is art - amateur radio is both a practical skill and an art too!

There are 3 levels of licenses currently available from the FCC - **Technician, General and Amateur Extra.**

Did you know that HamRadioPrep.com also offers a General Course? If you would like to study for your Technician and General at the same time use coupon code: **AMAZONSTUDY for 40% off!**

Talking On Ham Radio

Once you pass your technician exam, you will be assigned a Call Sign - that is the basic code you will use to identify yourself.

An amateur station is required to transmit its call-sign **at least every 10 minutes during and at the end of a communication.**

Call sign every <u>10 minutes</u> and <u>at the end of a communication</u>

What does a call sign look like, anyway? You've probably already seen them before. **K1XXX** is a valid call sign. Many operators choose to get custom call signs, just like vanity license plates. These are called vanity call signs.

Remember, **any licensed amateur** can get a vanity call sign - so feel free to choose a vanity call sign once you pass your exam!

When using tactical identifiers such as "Race Headquarters", you must transmit your call sign **at the end of each communication and every ten minutes during a communication**. (Hint - this doesn't change when operating in a competition or drill. <u>10 minutes</u>, <u>end of communication</u>).

You may decide to use your ham radio license to control remote control planes or other craft - this is one exception when you do not have to identify yourself.

When transmitting signals to control model craft, amateurs do not need to identify themselves. That would be no point in saying your call sign to a model plane anyway!

Good news English speakers! **The English language** is the only acceptable language to use for station identification when operating in a phone sub-band.

The method of call sign identification required for transmitting phone signals is to **send the call sign using CW OR phone emission.**

When using a self-assigned indicator, stroke, slant, or slash all have the same meaning! Example, you can say:
- **KL7CC stroke W3**
- **KL7CC slant W3**
- **KL7CC slash W3**

They are all correct!

In regards to any foul language - **any such language is prohibited**! Plain and simple, don't overthink it.

Broadcasting means **transmissions intended for reception by the general public** and is generally prohibited.

However, **when communications directly relate to the immediate safety of human life or protection of property**, an amateur station is authorized to transmit signals related to broadcasting, program production or news gathering when no other means is available.

An amateur radio station can make one-way transmissions **when transmitting code practice, information bulletins, or transmissions necessary to provide emergency communications.**

A Technician class licensee can be the control operator of a station operating in an exclusive Amateur Extra class operator segment **at no time.**

The FCC is In Charge

It's important to know who regulates and enforces the rules of the Amateur Radio Service in the United States as well – that is **the FCC.**

Your license is good for **10 years**! If your license expires, there is a grace period of **2 years** to renew it.

During this grace period, **no transmitting is allowed until the FCC license database shows that the license has been renewed.**

The FCC says that when identifying your station **it is encouraged** to use the phonetic alphabet. Instead of just saying the letter "A", for example, you could say "Alpha." This just makes it easier for the other station to understand you without mistakes.

Phonetic Alphabet			
A	Alpha	N	November
B	Bravo	O	Oscar
C	Charlie	P	Papa
D	Delta	Q	Quebec
E	Echo	R	Romeo
F	Foxtrot	S	Sierra
G	Golf	T	Tango
H	Hotel	U	Uniform
I	India	V	Victor
J	Juliet	W	Whiskey
K	Kilo	X	X-ray
L	Lima	Y	Yankee
M	Mike	Z	Zulu

Once you pass your technician exam, you will get your operator/primary station license. Each person can only have **one** of these. You can prove you have your license because **the control operator's operator/primary station license must appear in the FCC ULS consolidated licensee database** (Hint* keyword here is **database**).

You will know immediately after the exam if you passed, but you cannot start operating until your operator/station license grant appears in the FCC's license database, you are free to start operating.

As an amateur radio operator, you must keep your address up to date with the FCC. If the FCC cannot reach you at your mailing address, you could be subject to **revocation of the station license or suspension of the operator license.**

The FCC sets all of the rules for Amateur Radio in Part 97 of the Code of Federal Regulations - you can find it free online. Part 97 makes some important definitions:
- A beacon is **an amateur station transmitting communications for the purposes of observing propagation or related experimental activities.**
- A space station is **an amateur station located more than 50km above the Earth's surface.**

According to the FCC, willful interference to other amateur radio stations is permitted **at no time.**

If a repeater retransmits communications that violate FCC rules, **the control operator of the originating station** is responsible. (Hint, essentially its the old saying...you break it...you buy it!)

The originating station is responsible, so follow the rules when using a repeater!

When the control operator is not the station licensee, **the control operator and the station licensee are equally responsible.**

If you want to start a club, you can get a club station license grant. However, **a club must have at least 4 members** for the issuance of a club station license grant.

Third Party Communications means **a message from a control operator to another amateur station control operator on behalf of another person.**

You cannot use amateur radio to sell equipment commercially, but you may offer equipment for sale or trade **when the equipment is normally used in an amateur station and such activity is not conducted on a regular basis.**

When the communication is incidental to classroom instruction at an educational institution, the amateur operator may receive compensation.

Station Control

An amateur station is **never** permitted to transmit without a control operator.
The station licensee designates the station control operator.

Unless otherwise stated, **the station licensee** is assumed to be the control operator.

It is possible for another person to take control of the station to transmit. In that case, **the class of operator license held by the control operator** determine the privileges of the station.

The station licensee must make the station and its records available for FCC inspection **at any time upon request by an FCC representative.**
(Hint - don't overthink! If the FCC asks for your papers, hand them over!)

An amateur station control point is **the location at which the control operator function is performed.**

Operating In Other Countries and at Sea

FCC is in charge in the USA, and the ITU guides international communication.
The International Telecommunications Union (ITU) is **a United Nations agency for information and communication technology issues.**

You can even operate your amateur station in a foreign country **when the foreign country authorizes it.**

You cannot communicate with **any country whose administration has notified the International Telecommunications Union (ITU) that it objects to such communications.**

When talking internationally (to another country), an FCC-licensed amateur radio station is permitted to make **communications incidental to the purposes of the Amateur Radio Service and remarks of a personal character.**

For example, you may greet other amateurs and make small-talk. That would be "remarks of a personal character".

You may also transmit from **any vessel or craft located in international waters and documented or registered in the United States.**

A non-licensed person operating the station can speak to a foreign station **if the foreign station is one with which the U.S. has a third-party agreement.**

Emergency

RACES is the Radio Amateur Civil Emergency services. They are an organization of many functions:
- **A radio service using amateur frequencies for emergency management or civil defense communications**
- **A radio service using amateur stations for emergency management or civil defense communications**
- **An emergency service using amateur operators certified by a civil defense organization as being enrolled in that organization**

Space Stations

Any amateur whose license privileges allow them to transmit on the satellite uplink frequency can talk to a space station/satellite.

You can even make contact with the International Space Station (ISS) using 2 meter and 70cm band frequencies (**Any amateur holding a Technician or high-class license**) - that will be you!

You can only transmit encoded messages **when transmitting control commands to space stations or radio control craft**. You can also transmit music **incidental to an authorized retransmission of manned spacecraft communications.**

Repeaters

A Repeater station simultaneously retransmits the signals of another amateur station on a different channel or channels. It is used to repeat your signal and can help you communicate further. (Hint: A Repeater is the one that Retransmits.)

Repeater operation is an example of automatic control - it is automatic and doesn't need anyone operating! But it's not the only type of station that can retransmit.

Repeater, auxiliary, or space stations can automatically retransmit the signals of other amateur stations. (Hint, this answer is only one of two that contains space station in the answer)

Remote Control

When the FCC talks about Remote control in the Part 97 rules, they are talking about operating your station from afar. It's NOT about remote control craft!

Operating the station over the internet is an example of remote control as defined by Part 97 (Hint, don't get tricked by the remote control aircraft/boat answer!)

HamRadioPrep.com

An example of remote control in Part 97 is operating your station over the internet

The following are true about remote control operation:
- **The control operator must be at the control point**
- **The operator must be present at all times**
- **The control operator indirectly manipulates the control.**

Frequency and Band Basics

If you are looking for a recommendation on transmit/receive channels and other parameters for auxiliary and repeater stations, look to the **Volunteer Frequency Coordinator recognized by local amateurs.** (Hint* this is just a fancy way to say that the local Volunteer organization can recommend what frequencies to use!)

The Amateur operators in a local or regional area can select their own Frequency Coordinator to be in charge.

Here are two common bands:
- The 6-meter amateur band is **52.525Mhz**. (Hint, think of counting backward from 6....6, 52.525!)
- 146.52Mhz is the same as the **2-meter band.**

We will discuss frequencies in depth later – this is a quick one to memorize.

Some frequencies have limitations. With your Technician license, you will have phone privileges on the **10-meter band**! **The 10-meter band** is the only HF band available to amateurs for RTTY and data transmissions. (Hint* – when in doubt, answer 10-meter band!)

*****One of the writers of this course wore the number 10 in high school and 10m is one of his favorite bands!*

The frequencies between 219 and 220 Mhz are limited for **fixed digital message forwarding systems only.**

219 and 220MHz – Digital message forwarding system

50.0 MHz to 50.1 MHz and 144.0 MHz to 144.1 MHz are limited to CW only.

CW (Morse Code): 50.1MHz to 50.1MHz, 144.0MHz to 144.1 MHz

Once you get your license, remember not to interfere with other bands that could be used for non-amateur (such as military) communication. When using most frequencies (such as the 70cm band), **U.S. amateurs may find non-amateur stations in those portions and must avoid interfering with them.**

There are many reasons (Hint* - **all of the above)** not to set your transmit frequency exactly at the edge of an amateur band or sub-band:
- **You could end up interfering due to calibration error**
- **Modulation sidebands**
- **Transmitter frequency drift.**

There maximum power output in HF bands for Technician class operators is **200 watts**. For frequencies above 30 MHz, this extends all the way to **1500 watts.** (Hint* For this question always select the BIGGEST number..think of it like the peak of a mountain which is the top!)

Cracking the Technician License Exam

Chapter 1 Quiz

1) Who selects a Frequency Coordinator?

A. The FCC Office of Spectrum Management and Coordination Policy
B. The local chapter of the Office of National Council of Independent Frequency Coordinators
C. Amateur operators in a local or regional area whose stations are eligible to be a repeater or auxiliary stations
D. FCC Regional Field Office

2) Which of the following describes the Radio Amateur Civil Emergency Service (RACES)?

A. A radio service using amateur frequencies for emergency management or civil defense communications
B. A radio service using amateur stations for emergency management or civil defense communications
C. An emergency service using amateur operators certified by a civil defense organization as being enrolled in that organization
D. All of these choices are correct

3) When is willful interference to other amateur radio stations permitted?

A. To stop another amateur station which is breaking the FCC rules
B. At no time
C. When making short test transmissions
D. At any time, stations in the Amateur Radio Service are not protected from willful interference

4) Which of the following HF bands have frequencies available to the Technician class operator for RTTY and data transmissions?

A. 10 meter, 12 meter, 17 meter, and 40 meter bands
B. 10 meter, 15 meter, 40 meter, and 80 meter bands
C. 30 meter band only
D. 10 meter band only

5) What is the maximum peak envelope power output for Technician class operators using their assigned portions of the HF bands?

A. 200 watts
B. 100 watts
C. 50 watts
D. 10 watts

6) Except for some specific restrictions, what is the maximum peak envelope power output for Technician class operators using frequencies above 30 MHz?

A. 50 watts
B. 100 watts
C. 500 watts
D. 1500 watts

7) If your license has expired and is still within the allowable grace period, may you continue to operate a transmitter on Amateur Radio Service frequencies?

A. No, transmitting is not allowed until the FCC license database shows that the license has been renewed
B. Yes, but only if you identify using the suffix GP
C. Yes, but only during authorized nets
D. Yes, for up to two years

8) When may an amateur station transmit without on-the-air identification?

A. When the transmissions are of a brief nature to make station adjustments
B. When the transmissions are unmodulated
C. When the transmitted power level is below 1 watt
D. When transmitting signals to control a model craft

9) Who does the FCC presume to be the control operator of an amateur station, unless documentation to the contrary is in the station records?

A. The station custodian
B. The third-party participant
C. The person operating the station equipment
D. The station licensee

10) Which of the following is a requirement for the issuance of a club station license grant?

A. The trustee must have an Amateur Extra Class operator license grant
B. The club must have at least four members
C. The club must be registered with the American Radio Relay League
D. All of these choices are correct

CHAPTER 2: STATION OPERATION

Like we talked about in the first lesson, follow the rules. When can you break the rules?

Never, FCC rules always apply.

Let's start with one of the basics of station operation - calling another station. Always say the name of the other station first to let them know you're contacting them, then say yours.

Example: whisky-three-tango-alpha-bravo (W3TAB) this is kilo-one-foxtrot-delta-romeo (K1FDR) do you copy?

To call another station if you know the station's call sign, you **say the station's call sign, then identify with your call sign.**

Calling CQ means **calling any station.** This is a great way to make your first contact once you get your license!

You want to do some checks before calling "CQ, CQ." When choosing an operating frequency for calling CQ, you should:
- **Listen first to be sure that no one else is using the frequency**
- **Ask if the frequency is in use**
- **Make sure you are in your assigned band**

Just some common decency really!

To respond to station calling CQ, it's exactly the same. **Transmit the other station's call sign followed by your call sign.**

When making on-the-air test transmissions, it is required to **identify the transmitting station.**

Identify the transmitting station - say your call sign!

Simplex is when an amateur station is transmitting and receiving on the same frequency. (Hint* Simplex - simple!) Simplex are a great simple way to communicate.

Simplex channels are designated in the VHF/UHF band plans **so that stations within mutual communications range can communicate without tying up a repeater.**

The national calling frequency for FM simplex operations in the **2-meter band is 146.520 MHz** (Hint* The answer is the only one that has a 2!) Another important band is 70cm, more on that later.

SSB phone (Single Sideband phone) may be used in amateur bands above 50MHz and **is allowed in at least some portions of these bands.**

A band plan, beyond the privileges established by the FCC, is a **voluntary guide for using different modes or activities within an**

amateur band. (Hint: It's voluntary for a reason so don't be confused by the other answers that say its mandated by a club!)

Make sure you do not **talk too loudly** - it could cause your transmissions to break up on voice peaks.

When two stations transmitting on the same frequency interfere with each other, **common courtesy should prevail, but no one has an absolute right to an amateur frequency.**

Phonetic Alphabet

That cool way you always hear radio operators talk - Kilo-Mike-Xray...

That is called the phonetic alphabet. If you are transmitting and want to make sure that voice messages containing unusual words are received correctly, you can **spell them using the standard phonetic alphabet.**

If someone wants to be sure they were heard clearly, they may ask you for a "check". A "check" is **the number of words or word equivalents in the text portion of the message.**

Q Signals
Q signals are standard communication. Always three numbers. Always begin with Q. There are only a two of them on the test.

QRM	You are receiving interference from other stations.
QSY	Indicates that you are changing frequency (Hint* QSY - frequency ends in y)

Repeaters

Repeaters just take a radio signal in and repeat it – if you talk to a repeater, it will take your signal and broadcast it out further.

To talk on a repeater, you only need to say **your call sign** which indicates that you are listening on a repeater and looking for a contact. (Hint: Don't get tricked by saying having to say anything else!)

There is a little setup required before talking on a repeater. You have to set your radio to match the repeater's offset.

"Repeater offset" is **the difference between a repeater's transmit frequency and its receive frequency.**

You also may need to setup squelch.

A sub-audible tone transmitted along with normal voice audio to open the squelch of a receiver is **CTCSS**. Think of this as sort of a 'key' that allows you to use a repeater.

***CTCSS** - sub-audible tone to open the squelch of the receiver*

If you cannot access a repeater whose output you can hear, it could be because of:

- **Improper transceiver offset**
- **The repeater may require a proper CTCSS tone from your transceiver**
- **The repeater may require a proper DCS tone from your transceiver**

If a station is not strong enough to keep a repeater's receiver squelch open, you might be able to receive the station's signal by **listening on the repeater's input frequency.** Basically, try to find the original signal for the repeater. The most common use of the "reverse split" function of a VHF/UHF transceiver is also **to listen on a repeater's input frequency.**

The 2-meter band is very important for amateur radio operators because its local and reliable! Let's say you're talking on 2-meter and want to hit a repeater to get your signal out. You have to match the repeater's offset frequency. **Plus or minus 600kHz** is a common repeater frequency offset in the **2-meter band.**

For **70cm,** a common repeater frequency offset is **plus or minus 5MHz.**

*70cm repeater - **plus or minus 5MHz** offset is common*

A linked repeater network is **a network of repeaters where signals received by one repeater are repeated by all the repeaters.**

DTMF tones are used to control repeaters linked by the Internet Relay Linking Project (IRLP) protocol. (Hint: IRLP is a 4 syllable acronym just like DTMF; On the test DTMF is the only 4 syllable acronym!)

***DTMF** tones are used to control repeaters linked by IRLP*

A talk group on a DMR digital repeater is a **way for groups of users to share a channel at different times without being heard by other users on the channel.**

You can join a digital repeater's "talk group" by **programming your radio with the group's ID or code.**

Net Operations

The term "NCS" in net operations means **Net Control Station.**

The term "traffic" in net operation means **formal messages exchanged by net stations.**

Once you've checked into a net, **remain on frequency without transmitting until asked to do so by the net control station.**

RACES and ARES, Emergencies

If you're considering learning more about ham radio for emergencies RACES and ARES **both may provide communications during emergencies.**

The Amateur Radio Emergency Service (ARES) is a group of **licensed amateurs who have voluntarily registered their qualifications and equipment for communications duty in the public service.**

When reporting an emergency, **begin your transmission by saying "Priority" or "Emergency"** to get the immediate attention of a net control system. (Hint* In an emergency, just say Emergency!)

In special situations involving the immediate safety of human life or protection of property, you may go out of your range as licensed as a Technician.

Good traffic handling means **passing messages exactly as received**. This is important in an emergency so there is no confusion!

Informal traffic messaging, **the information needed to track the message** is stored in the preamble.

Chapter 2 Quiz

1) What is the national calling frequency for FM simplex operations in the 2-meter band?

A. 146.520 MHz
B. 145.000 MHz
C. 432.100 MHz
D. 446.000 MHz

2) What is the most common use of the "reverse split" function of a VHF/UHF transceiver?

A. Reduce power output
B. Increase power output
C. Listen on a repeater's input frequency
D. Listen on a repeater's output frequency

3) Which of the following describes a linked repeater network?

A. A network of repeaters where signals received by one repeater is repeated by all the repeaters
B. A repeater with more than one receiver
C. Multiple repeaters with the same owner
D. A system of repeaters linked by APRS

4) When do the FCC rules NOT apply to the operation of an amateur station?

A. When operating a RACES station
B. When operating under special FEMA rules
C. When operating under special ARES rules
D. Never, FCC rules always apply

5) What information is contained in the preamble of a formal traffic message?

A. The email address of the originating station
B. The address of the intended recipient
C. The telephone number of the addressee
D. The information needed to track the message

6) What is meant by the term "check," in reference to a formal traffic message?

A. The number of words or word equivalents in the text portion of the message
B. The value of a money order attached to the message
C. A list of stations that have relayed the message
D. A box on the message form that indicates that the message was received and/or relayed

7) What is the Amateur Radio Emergency Service (ARES)?

A. Licensed amateurs who have voluntarily registered their qualifications and equipment for communications duty in the public service
B. Licensed amateurs who are members of the military and who voluntarily agreed to provide message handling services in the case of an emergency
C. A training program that provides licensing courses for those interested in obtaining an amateur license to use during emergencies
D. A training program that certifies amateur operators for membership in the Radio Amateur Civil Emergency Service

8) Which of the following is an accepted practice for an amateur operator who has checked into a net?

A. Provided that the frequency is quiet, announce the station call sign and location every 5 minutes
B. Move 5 kHz away from the net's frequency and use high power to ask other hams to keep clear of the net frequency
C. Remain on frequency without transmitting until asked to do so by the net control station
D. All of these choices are correct

9) Which of the following is a characteristic of good traffic handling?

A. Passing messages exactly as received
B. Making decisions as to whether messages are worthy of relay or delivery
C. Ensuring that any newsworthy messages are relayed to the news media
D. All of these choices are correct

10) Are amateur station control operators ever permitted to operate outside the frequency privileges of their license class?

A. No
B. Yes, but only when part of a FEMA emergency plan
C. Yes, but only when part of a RACES emergency plan
D. Yes, but only if necessary in situations involving the immediate safety of human life or protection of property

CHAPTER 3: RADIO WAVE CHARACTERISTICS

Basic Concepts

It's not magic that makes radio waves travel through the air - it's electromagnetic energy!

Electric and magnetic fields are the two components of a radio wave.

When we combine electric and magnetic we get electro + magnetic = electromagnetic waves. **Electromagnetic waves** carry signals between transmitting and receiving sessions.

Have you ever thought to yourself, it Radio waves travel at the **speed of light**. The speed of **light is 300,000,000 meters per second**. Hint* the answer is the biggest number!

Polarization

The orientation of the electric field is used to describe the polarization of a radio wave.

Remember, it's the ELECTRIC field that describes the polarization of a radio wave! The magnetic field is actually perpendicular.

Depending on how you setup your antenna, it could be horizontally polarized or vertically polarized.

Horizontal polarization is normally used for long-distance weak-signal CW and SSB contacts using the VHF and UHF bands.

Line of Sight Communication

The most simple way to talk to another station is just a straight signal from your radio to theirs. This is called line of sight. Line of site can only talk as far as the radio horizon.

The radio horizon is **the distance over which two stations can communicate by a direct path.**

For line of sight, you want to make sure the antennas are polarized the same!

If the antennas at opposite ends of a VHF or UHF line of sight radio link are not using the same polarization, **signals could be significantly weaker.**

Radio waves can even sometimes penetrate buildings.

UHF signals are more effective than VHF signals in buildings because **the shorter wavelength allows them to more easily penetrate the structure of buildings.**

Surprisingly, even without reflecting, radio waves can communicate a little bit further than a line of sight!
The Earth seems less curved to radio waves to light, so VHF and UHF radio signals usually travel even further than visual line of sight!

Using Reflection to Communicate Even Further

Let's say you're trying to talk to a repeater but there is a building in the way. Using a directional antenna, you may be able to access a distant repeater even if you have an obstruction by **finding a path that reflects signals to the repeater.**

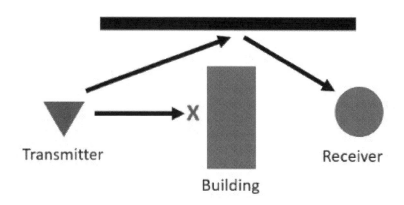

You could bounce radio waves off a mountain or some obstacle. This is called **knife-edge diffraction** can cause radio signals to be heard despite obstructions between the two stations.

You want to avoid reflecting off of multiple objects. If data signals propagate over multiple paths, **error rates are likely to increase.**

But for really long distance communication you've got to think bigger than buildings or obstacles - you can also reflect off the atmosphere. This is called "skipping" signals off the atmosphere and it enables long distance communication.

The **ionosphere** is the part of the atmosphere that enables the propagation of radio signals around the world. The ionosphere has two layers important for radio, the E-layer and F-layer.

You can you use both vertically polarized or horizontally polarized antennas when skipping signals off the ionosphere.
Because the ionosphere elliptically polarizes skip signals, **either vertically or horizontally polarized antennas may be used for transmission or reception.**

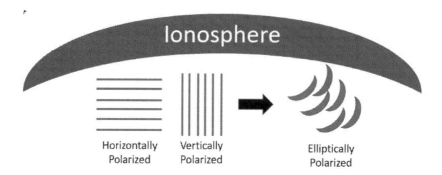

Vertical and horizontally polarized antennas both transmit and receive skip signals from the ionosphere because the ionosphere polarizes them elliptically!

VHF signals can be **refracted from a sporadic E layer** to receive over long distances.

Sporadic E propagation is commonly associated with occasional strong over-the-horizon signals on the 10, 6, and 2-meter bands.

There are limitations especially with UHF for skipping signals. UHF signals are rarely heard from stations outside the local coverage area **because UHF signals are usually not reflected by the ionosphere.**

The troposphere is below the ionosphere but is still very useful to communicate over long distances.

Tropospheric scatter can be used by both VHF and UHF to communicate up to 300 miles on a regular basis!

Long distance communication can be facilitated by a really interesting effect called tropospheric ducting. Tropospheric ducting is caused by **temperature inversions in the atmosphere** and can help carry your signal!

Wavelength and Frequency

The distance a radio wave travels during one complete cycle is the **Wavelength**.

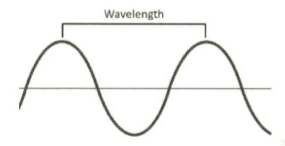

You will always hear HAMs talking about Wavelength and Frequency. They are closely related. **The wavelength gets shorter as the frequency increases.**

Wavelength in meters equals 300 divided by frequency in megahertz.

Remember, the formula to calculate wavelength in meters is:

The approximate wavelength is often used to identify the different frequency bands.

There are three general levels of frequency. If you can remember what the letters stand for, it is easy to remember which is the highest!

In order of lowest to the highest frequency they are:

HF	**High** Frequency	3 to 30 MHz
VHF	**Very High** Frequency	30 to 300 MHz
UHF	**Ultra High** Frequency	300 to 3000 MHz

High Frequency is from 3 to 30 MHZ

Don't get confused by HF. This is the LOWEST FREQUENCY of the three ranges! Very High Frequency and Ultra High Frequency are even higher!

Surprisingly the low frequency waves of HF are better for long distance communication. Low frequency means long distance in this case!

Sunspots and Meteor Scatter

Sunspots can hamper or improve radio propagation.

From dawn to shortly after sunset during periods of high sunspot activity is the best time for long-distance 10 meter band propagation via the F-layer.

Six or ten-meter bands provide long distance communications during the peak of the sunspot cycle.

The 6-meter band is best suited for communicating via meteor scatter.

Distortion

The rapid fluttering sound heard from mobile stations that are moving are sometimes called **picket fencing.**

VHF signals can also reflect off of the aurora - in this case, **the signals exhibit rapid fluctuations of strength and often sound distorted.**

The **random combining of signals arriving via different paths** can cause irregular fading of signals received by ionospheric reflection.

If another operator reports that your station's two-meter signals were strong just a moment ago, but now they are weak or distorted, **try moving a few feet or changing the direction of your antenna if possible, as reflections may be causing multi-path distortion.**

Chapter 3 Quiz

1) What should you do if another operator reports that your station's 2-meter signals were strong just a moment ago, but now they are weak or distorted?

A. Change the batteries in your radio to a different type
B. Turn on the CTCSS tone
C. Ask the other operator to adjust his squelch control
D. Try moving a few feet or changing the direction of your antenna if possible, as reflections may be causing multi-path distortion

2) Why might the range of VHF and UHF signals be greater in the winter?

A. Less ionospheric absorption
B. Less absorption by vegetation
C. Less solar activity
D. Less tropospheric absorption

3) What antenna polarization is normally used for long-distance weak-signal CW and SSB contacts using the VHF and UHF bands?

A. Right-hand circular
B. Left-hand circular
C. Horizontal
D. Vertical

4) What is the name for the distance a radio wave travels during one complete cycle?

A. Wave speed
B. Waveform
C. Wavelength
D. Wave spread

5) What property of a radio wave is used to describe its polarization?

A. The orientation of the electric field
B. The orientation of the magnetic field
C. The ratio of the energy in the magnetic field to the energy in the electric field
D. The ratio of the velocity to the wavelength

6) What are the two components of a radio wave?

A. AC and DC
B. Voltage and current
C. Electric and magnetic fields
D. Ionizing and non-ionizing radiation

7) Why are direct (not via a repeater) UHF signals rarely heard from stations outside your local coverage area?

A. They are too weak to go very far
B. FCC regulations prohibit them from going more than 50 miles
C. UHF signals are usually not reflected by the ionosphere
D. UHF signals are absorbed by the ionospheric D layer

8) Which of the following is an advantage of HF vs VHF and higher frequencies?

A. HF antennas are generally smaller
B. HF accommodates wider bandwidth signals
C. Long distance ionospheric propagation is far more common on HF
D. There is less atmospheric interference (static) on HF

9) What is a characteristic of VHF signals received via auroral reflection?

A. Signals from distances of 10,000 or more miles are common
B. The signals exhibit rapid fluctuations of strength and often sound distorted
C. These types of signals occur only during winter nighttime hours
D. These types of signals are generally strongest when your antenna is aimed west

10) Which of the following propagation types is most commonly associated with occasional strong over-the-horizon signals on the 10, 6, and 2-meter bands?

A. Backscatter
B. Sporadic E
C. D layer absorption
D. Gray-line propagation

CHAPTER 4: AMATEUR RADIO PRACTICES

Station Setup

To have a station, you need to power it. In order to determine the minimum current capacity needed for a transceiver power supply, you must consider the:

- **Efficiency of the transmitter at full power output**
- **Receive and control circuit power**
- **Power supply regulation and heat dissipation**

Efficiency of the transmitter at full power output

Receive and control circuit power

Power supply regulation and heat dissipation

*Consider **all of the above** when determining the minimum current capacity for a transceiver power supply*

To enable quick access to a favorite frequency you can **store the frequency in a memory channel.**

To enter the operating frequency on a modern transceiver, you can use **the keypad or VFO knob.**

The scanning function of a transceiver is used **to scan through a range of frequencies to check for activity.**

"RIT" stands for **Receiver Incremental Tuning.** Using RIT you can tune your receiver without changing your transmit. If the voice pitch of a single-sideband signal seems too high or low, you could use **the**

receiver RIT or clarifier to change the frequency. (Hint: It's the only answer with a tune!)

Receiver Incremental Tuning

Use the receiver RIT or clarifier if if the voice pitch seems too high or low

---

Wiring

Wiring between the power source and radio should be heavy-gauge wire and kept as short as possible **to avoid voltage falling below that needed for proper operation.**

The proper location for an external SWR meter is **in series with the feed line, between the transmitter and antenna.**

The negative return connection of a mobile transceivers power cable should be connected **at the battery or engine block ground strap.**

A **Flat strap** provides the lowest impedance to RF signals.

Noise and Distortion

Noise and distortion is all around us. Your car can even cause interference.

The alternator in your car can cause a high-pitched whine that varies with engine speed in a mobile transceiver's receive audio.

To get rid of power line noise or ignition noise, you can use a **noise blanker**.

Turn on the noise blanker to reduce ignition interference to a receiver.

A **ferrite choke** can be used to cure distorted audio caused by RF current on the shield of a microphone cable.

If a transmitter is operated with the microphone gain too high, **the output signal might become distorted.**

Squelch control is used **to mute receiver output noise when no signal is being received.**

Automatic Gain Control, or (AGC) is used **to keep received audio relatively constant.**

Filters are important for cleaning up noisy signals. Choosing the right bandwidth for your filter depends on what mode you're talking in.
- **2400 Hz** is an appropriate receive filter bandwidth for minimizing noise and interference for **SSB** reception.

- **500 Hz** is an appropriate receive filter bandwidth for minimizing noise and interference for **CW** reception.

In other words, if you want to add a filter for noise and interference for **SSB reception**, you can choose a **2400 Hz** bandwidth filter.

If you want to add a filter for **CW**, it may be wise to get a **500 Hz** bandwidth filter.

(Hint: CW, or morse code, uses very low bandwidth because it is just tones! You only need a 500Hz bandwidth filter.)

Some radios have receive bandwidth choices so you can change this filter depending on if you are using morse-code, SSB voice, or other!

Having multiple receive bandwidth choices on a multimode transceiver **permits noise or interference reduction by selecting a bandwidth matching to the mode.**

(Hint: Don't be fooled by the answer saying you can listen to multiple modes at once. You just want to listen to one mode, like SSB phone or CW, but you can select the right bandwidth for the filter!)

Computers

A computer can be used as part of an amateur station in many ways.
- **For logging contacts and contact information**
- **For sending and/or receiving CW**
- **For generating and decoding digital signals**

The microphone or line input is the computer sound card port that is connected to a transceivers headphone or speaker output for operating digital modes. (Hint: Don't fall for the answer that says headphone. The correct answer contains the word microphone!)

The sound card provides audio to the radio's microphone input and converts received audio to digital form.

The following connections might be used between a voice transceiver and a computer for digital operation are:
- **Receive audio**
- **Transmit audio**
- **Push to Talk (PTT)**

Cracking the Technician License Exam

All of these answers are correct!

Chapter 4 Quiz

1) Which of the following is an appropriate receive filter bandwidth for minimizing noise and interference for SSB reception?

A. 500 Hz
B. 1000 Hz
C. 2400 Hz
D. 5000 Hz

2) Which of the following is an appropriate receive filter bandwidth for minimizing noise and interference for CW reception?

A. 500 Hz
B. 1000 Hz
C. 2400 Hz
D. 5000 Hz

3) What is the function of automatic gain control, or AGC?

A. To keep received audio relatively constant
B. To protect an antenna from lightning
C. To eliminate RF on the station cabling
D. An asymmetric goniometer control used for antenna matching

4) Which of the following could be used to remove power line noise or ignition noise?

A. Squelch
B. Noise blanker
C. Notch filter
D. All of these choices are correct

5) Which of the following is a use for the scanning function of an FM transceiver?

A. To check incoming signal deviation
B. To prevent interference to nearby repeaters
C. To scan through a range of frequencies to check for activity
D. To check for messages left on a digital bulletin board

6) What must be considered to determine the minimum current capacity needed for a transceiver power supply?

A. Efficiency of the transmitter at full power output
B. Receiver and control circuit power
C. Power supply regulation and heat dissipation
D. All of these choices are correct

7) How might a computer be used as part of an amateur radio station?

A. For logging contacts and contact information
B. For sending and/or receiving CW
C. For generating and decoding digital signals
D. All of these choices are correct

8) Why should wiring between the power source and radio be heavy-gauge wire and kept as short as possible?

A. To avoid voltage falling below that needed for proper operation
B. To provide a good counterpoise for the antenna
C. To avoid RF interference
D. All of these choices are correct

9) Which computer sound card port is connected to a transceiver's headphone or speaker output for operating digital modes?

A. Headphone output
B. Mute
C. Microphone or line input
D. PCI or SDI

10) What is the proper location for an external SWR meter?

A. In series with the feed line, between the transmitter and antenna
B. In series with the station's ground
C. In parallel with the push-to-talk line and the antenna
D. In series with the power supply cable, as close as possible to the radio

CHAPTER 5: ELECTRICAL PRINCIPLES

First Concepts

There are a few main electronic concepts you need to read first:
- **Voltage** is the electromotive force (EMF) that causes electrons to flow. It is measured in **Volts**.
    - Hint* voltage is an electromotive force – a force that pushes the current.
- **Current** is the flow of electrical charge. Electrical current is measured in **Amperes**.
    - Hint* – Think of current like the flow of water in a pipe. The voltage (pressure) pushes the current!
- **Resistance** is the property that resists current flow.
    - Hint* think of rocks or obstructions blocking the flow of water in the pipe. That's our resistance.
- **Electrical Power** – electrical power is measured in **Watts**
    - Hint* If you like football, you can think of JJ Watt – a big and powerful football player!
    - **Power** is the **rate at which electrical energy is used**.
    - Hint* if you use a lot of power, you will run out of energy!

AC and DC

Have you ever thought about how the AC power from the plugs in your house is different than the DC current from a battery? If you have, you probably already know some answers!

The most simple kind of voltage is what you find in a battery – **Direct Current**. Direct Current from a battery is simple and **flows in one direction**.

DC batteries are great for mobile stations, which usually use standard DC voltages. A mobile transceiver usually requires **about 12 volts**. Don't overthink this question! 12V is one of the most common battery voltages.

What about the voltage from the outlet in your wall? This is called AC or **Alternating Current**. The current is alternating because it **changes direction regularly**. We can measure the rate that the current changes direction per second, and that is called the **frequency**.

Conductors and Insulators

In electrical theory, there are insulators and conductors.

Copper is a good conductor. Hint* they both begin with C! Just think of any Copper wire.

And what do you put around a conductor, to protect from the charge? Just like how a coat could insulate you from the cold, an insulator in electrical circuits protects from the electrical charge. **Glass is a good insulator**.

Math and Units

You will also need to know a bit of math for the test. If this isn't your favorite section, don't sweat it! The test will **only include 4 math questions.**

How do you convert 1.5 Amperes to milliamperes?

There are 1000 milliamperes in one ampere. To change 1.5 amperes to milliamperes, we just need to move the decimal place three times:

1.5 ampere = 1500 milliamperes.

We can use the same method to convert 1,500,000 hertz to kHz, but in the opposite direction!

1,500,000Hz = 1,500kHz

Now you can try it without the graphic. How many volts are equal to one kilovolt? A kilovolt is bigger than a volt, so we need to move the decimal three places to make a kilovolt.

1,000V = 1kV

The proper abbreviation for megahertz is **MHz.**

To change a volt to microvolt, we change the decimal in the opposite direction:

One one-thousandth of a volt = One one-millionth of a volt

Now you're getting it! Let's try one with watts.

How many watts are equivalent to 500 milliwatts?

It doesn't make a difference if we are talking about watts, volts, or amps, the concept is the same!

500mW = .5 Watts (move the decimal place by three!).

In summary,

Kilo -> move the decimal 3 places to the right (larger)

Milli -> move the decimal 3 places to the left (smaller)

Billion	Million	Thousand	1	One thousandth	One millionth	One billionth	One trillionth
Giga	Mega	Kili	Base Unit	Mili	Micro	Nano	Pico
< 3 Decimals >	< 3 Decimals >	< 3 Decimals >	< 3 Decimals >	< 3 Decimals >	< 3 Decimals >	< 3 Decimals >	< 3 Decimals >

Here is a tricky one.

A picofarad is 1/1000 of a microfarad. **1,000 microfarads** is the same as 1,000,000 picofarads.

Power is often measured in decibels (dB). Decibels is a scale of the amplitude of something. Someone may use a hearing aid to increase the decibels, or intensity, of the sound to their ear.

Decibel scale works differently.

Number in Decibels	Multiplier
3dB	Times 2
6dB	Times 4
10dB	Times 10

Double in decibel is 3dB. 5 watts to 10 watts is **3dB.**

10dB is a multiplier of times 10. An increase of 20 watts to 200 watts would be **10dB**.

What about a decrease in power? That would be negative decibels! A decrease of 12 watts to **3 watts is -6dB**.

Electric Principles

The ability to store energy in an electric field is **called Capacitance**, and it is measured in farads. Capacitors look like this, with the pluses and minuses added to show the electric field:

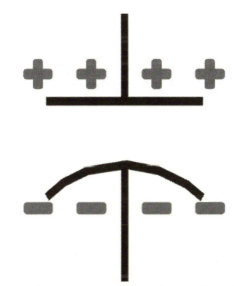

A capacitor stores energy in an electric field.

The ability to store energy in a magnetic field is called **inductance**. The basic unit of inductance is **the Henry**. Below is an inductor, with an oval around it that represents the magnetic field.

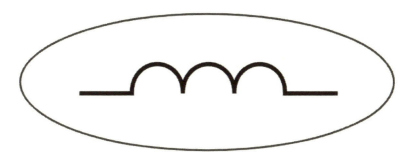

An inductor stores energy in a magnetic field

The basic unit of frequency is **Hertz**.

The basic unit of impedance is **Ohms**. Impedance is the **opposition to AC flow in a circuit**. Hint* Impedance is impeding the flow.

RF refers to radio frequency signals of all types. Hint* don't overthink it! RF is radio frequency!

Radio waves are made up of **electromagnetic energy.**

Formulas:

The most famous formula is Ohm's Law:

Voltage (E) equals current (I) multiplied by resistance (R).

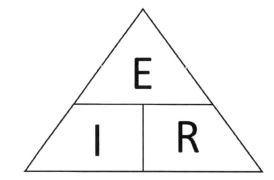

E (voltage) = current (I) times resistance (R)

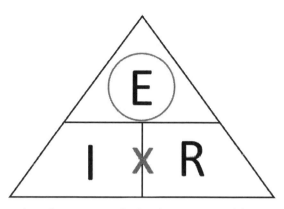

Using our triangle we can circle the current (I) and make the equation.

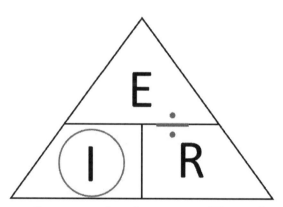

As you can see from our triangle, **current (I) equals voltage (E) divided by resistance (R).**

Let's try calculating resistance, and you can imagine your own triangle. Circle resistance. E is over the I, so resistance is voltage divided by current.

If we have 90 volts and 3 amperes, then the resistance is 90 / 3 = **30 ohms.**

Remember the triangle! The test will almost always ask one of these!

Let's take a look at the power equation now.

Power equals voltage times current. Hint* more voltage, more current, more power!

Now we can use this on some simple math. If we have 13.8 volts and 10 amps, then 13.8V times 10 amps = **138 watts**.

Series and Parallel

There are two ways to connect components in a circuit - series or parallel.

Series

When components are in series, they are connected in a line.

In a **series** circuit, the current is the same through all components because it has to pass through the same line **(it is unchanged)**.

The voltage between the components in series **is determined by the type and value of the components**.

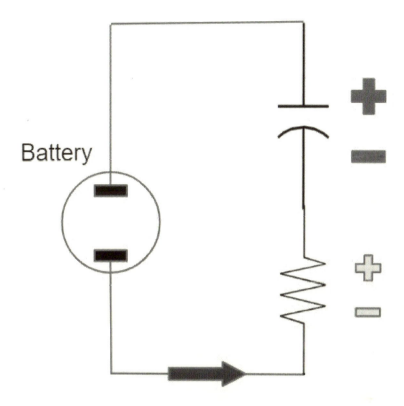

Current is running through these two components in series - there is only one path. The voltage is determined by the type and value of the components.

Parallel

In a **parallel** circuit, the voltage is the same across all components. Therefore, the voltage across two components in parallel with a voltage source is the **same voltage as the source**.

In parallel the current has two paths, it **divides between them dependent on the value of the components.**

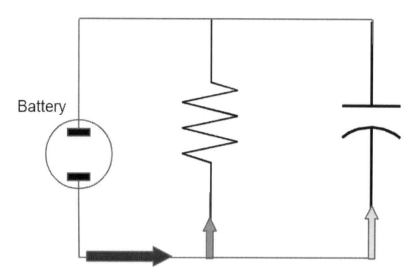

Look how the current from the battery (the different size arrows) divides between the the resistor and capacitor. It favors the easiest path, just like a river dividing into two streams. The current has two paths to choose, so the current divides between them depending on the value of the components.

Chapter 5 Quiz

1) Which term describes the rate at which electrical energy is used?

A. Resistance
B. Current
C. Power
D. Voltage

2) What happens to current at the junction of two components in parallel?

A. It divides between them dependent on the value of the components
B. It is the same in both components
C. Its value doubles
D. Its value is halved

3) What is the voltage across each of two components in series with a voltage source?

A. The same voltage as the source
B. Half the source voltage
C. It is determined by the type and value of the components
D. Twice the source voltage

4) What is the voltage across each of two components in parallel with a voltage source?

A. It is determined by the type and value of the components
B. Half the source voltage
C. Twice the source voltage
D. The same voltage as the source

5) What is impedance?

A. A measure of the opposition to AC current flow in a circuit
B. The inverse of resistance
C. The Q or Quality Factor of a component
D. The power handling capability of a component

6) What is a unit of impedance?

A. Volts
B. Amperes
C. Coulombs
D. Ohms

7) What is the proper abbreviation for megahertz?

A. mHz
B. mhZ
C. Mhm
D. MHz

8) What is the amount of change, measured in decibels (dB), of a power increase from 20 watts to 200 watts?

A. 10 dB
B. 12 dB
C. 18 dB
D. 28 dB

9) Which of the following frequencies is equal to 28,400 kHz?

A. 28.400 MHz
B. 2.800 MHz
C. 284.00 MHz
D. 28.400 kHz

10) If a frequency display shows a reading of 2425 MHz, what frequency is that in GHz?

A. 0.002425 GHz
B. 24.25 GHz
C. 2.425 GHz
D. 2425 GHz

CHAPTER 6: ELECTRICAL COMPONENTS

Intro

All circuits are made up of the same basic components:
- **Resistor:** opposes flow in a DC circuit. Hint* think resists flow
- **Potentiometer:** often used as an adjustable volume control. It is used to control the Resistance.
- **Capacitor:** stores energy in an electric field. It is two conductive surfaces separated by an insulator.
- **Inductor:** stores energy in a magnetic field. It is made from a coil of wire.
- **Switch:** used to connect or disconnect electrical circuits.
- **Fuse:** used to protect other circuit components from current overload.
- **Diode:** allows current to flow in only one direction (Hint* a diode looks like an arrow in a diagram)
    - A diode has a **cathode and an anode.**
    - **A stripe** indicates the cathode of a semiconductor diode.
    - **LED** is a Light Emitting Diode. Hint* think LED lights!

An inductor is made of a coil of wire that creates a magnetic field

You can combine a **capacitor** with an inductor to make a tuned circuit.

When you combine several of these components together it is called an **integrated circuit.**

It is common to use shielded wire in a circuit **to prevent coupling of unwanted signals to or from the wire.**

Batteries

Remember that **many battery types are rechargeable: Nickel-metal hydride, lithium-ion, and lead-acid gel-cell.**

Carbon-zinc is one battery type that is not rechargeable.

Transistors

Transistors have many uses and they will definitely come up on the test!

A Transistor uses a voltage or current signal to control current flow. Because they can control current flow with voltages or current, they are very common and have a lot of uses.

There are many types of the transistors, but the most common is the FET which stands for **Field Effect Transistor.**

A Transistor can be used as an electronic switch or amplifier. When it is used as an amplifier this is called gain.

A Transistor can amplify signals. That is why the Transistor can be the primary gain-producing components in an RF power amplifier.

A Transistor can be made of three layers of semiconductor material. (Hint* Transistors have a lot of uses. If in doubt, guess transistor).

Schematic Symbols

The test has 3 possible diagrams, and it will ask you about these symbols. Be prepared for a diagram question!

An electrical wiring diagram that uses the standard component symbols is called **a schematic**. Diagrams with these symbols show **the way the components are interconnected.**

Anytime you see the arrow pointing to the middle of the symbol, it is the variable version of that component.

Schematic Symbols

Component	Symbol	Component	Symbol
Resistor	—/\/\/—	Battery	(symbol)
Variable Resistor	(symbol)	Light Emitting Diode	(symbol)
Capacitor	—|(—	Transformer	(symbol)
Transistor	(symbol)	Variable Inductor	(symbol)
Switch (Hint* Single-pole single throw)	—o o—	Antenna	▽

A few more!

Let's talk about some more complex devices made up of these components.

A common one you can see on the telephone pole outside of your house is a transformer. Transformers are used to go from one AC voltage to another AC voltage. In ham radio, a **Transformer** is commonly used to change 120V AC house current to a lower AC voltage.

However, what if you need to go from an AC current to a DC current? A transformer can only change one AC signal to another, so that won't work! Use a rectifier. A **rectifier** changes alternating current into a varying direct current signal.

A relay is an **electrically-controlled switch.**

A **meter** can be used to display signal strength on a numeric scale. This one is a gimme! Just think about how a meter stick shows numbers!

A **regulator** circuit controls voltage from a power supply. (Hint* just think that it *regulates* the power.)

Chapter 6. Electrical Components

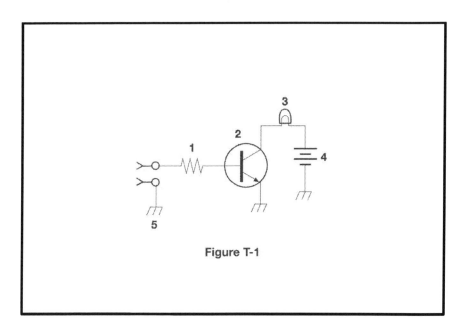

Figure T-1

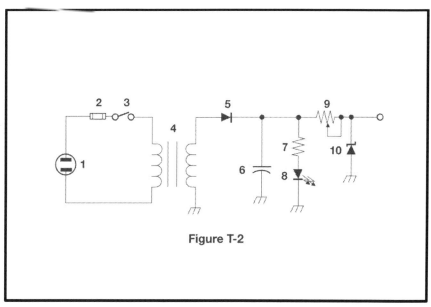

Figure T-2

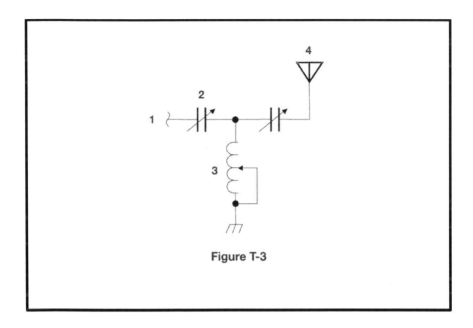

Figure T-3

1) What is component 1 in figure T1?

A. Resistor
B. Transistor
C. Battery
D. Connector

2) What is component 6 in figure T2?

A. Resistor
B. Capacitor
C. Regulator IC
D. Transistor

3) What is component 3 in figure T3?

A. Connector
B. Meter
C. Variable capacitor
D. Variable inductor

4) How is the cathode lead of a semiconductor diode often marked on the package?

A. With the word "cathode"
B. With a stripe
C. With the letter C
D. With the letter K

5) What does the abbreviation LED stand for?

A. Low Emission Diode
B. Light Emitting Diode
C. Liquid Emission Detector
D. Long Echo Delay

6) What does the abbreviation FET stand for?

A. Field Effect Transistor
B. Fast Electron Transistor
C. Free Electron Transmitter
D. Frequency Emission Transmitter

7) What electrical component is used to connect or disconnect electrical circuits?

A. Magnetron
B. Switch
C. Thermistor
D. All of these choices are correct

8) What electrical component is used to protect other circuit components from current overloads?

A. Fuse
B. Capacitor
C. Inductor
D. All of these choices are correct

9) Which of the following battery types is rechargeable?

A. Nickel-metal hydride
B. Lithium-ion
C. Lead-acid gel-cell
D. All of these choices are correct

10) Which of the following battery types is not rechargeable?

A. Nickel-cadmium
B. Carbon-zinc
C. Lead-acid
D. Lithium-ion

CHAPTER 7: STATION EQUIPMENT

Equipment

A radio has two basic functions - to transmit your signal to to other people's radios, and to receive signals from other radios to you. When you buy a radio, usually the transmitter and receiver are combined into one unit that does both - a transceiver. A transceiver is a **unit combining the functions of a transmitter and a receiver.**

Hand-held radios can easily switch between transmit and receive by using the PTT button. PTT means the **push-to-talk function that switches between receive and transmit**

An important part of transmitting is creating a signal that varies. There is a special kind of circuit for that. An **oscillator** is the name of a circuit that generates a signal at a specific frequency.

Amplifier Equipment

An **RF power amplifier** increases the low-power output from a handheld transceiver.

The SSB/CW-FM switch on a VHF power amplifier **sets the amplifier for the proper operation in the selected mode.**

(Hint* changing the switch on the power amplifier only sets the AMPLIFIER for the correct mode. It doesn't change the mode of the transmitter!)

An RF preamplifier is installed **between the antenna and the receiver.**

Devices for Converting Signals

A **transverter** converts the RF input and output of a transceiver to another band.

A **mixer** is used to convert a radio signal from one frequency to another.

Measurement Devices

To build and troubleshoot electronics, you are going to need to measure all kinds of things: voltage, current, resistance, power, and more. Let's talk about the devices you can use.

An **ohmmeter** is used to measure resistance.

(Hint* an ohm is a measurement of resistance)

What is probably happening when an ohmmeter, connected across an unpowered circuit, initially indicates a low resistance and then shows increasing resistance with time? This is a good sign that **the circuit contains a large capacitor.**

When measuring circuit resistance with an ohmmeter, **ensure that the circuit is not powered.**

An **ammeter** is used to measure electric current. (Hint* an amp is a measurement of current)

An ammeter should be connected **in series with the circuit.**

A **voltmeter** is used to measure electric potential or electromotive force. (Hint* a volt is a measure of electromotive force or voltage)

A voltmeter should be connected **in parallel with the circuit.**

When measuring with a voltmeter **ensure that the voltmeter and leads are rated for use at the voltages to be measured.**

A multimeter measures **voltage and resistance.**

Attempting to measure voltage when using the resistance setting could damage a multimeter.

Sometimes to test something, you may make a dummy load. A dummy load consists of a **non-inductive resistor and a heat sink.** The primary purpose of a dummy load is **to prevent transmitting signals over the air when making tests.**

An antenna analyzer can be used to determine if an antenna is resonant at the desired operating frequency.

Solder

Solder is used to "glue" wires and components in circuits. **Rosin-core solder** is best for radio and electronic use. Wait for it to get cold before touching. When it is cold it will have **a grainy or dull surface.**

SWR

There is important concept in ham radio for correctly setting up your antenna and radio called impedance matching. Essentially you need your transmitter impedance to match the antenna and load impedance. The Standing Wave Ratio, or SWR, is used to measure impedance matching.

SWR is **a measure of how well a load is matched to a transmission line.** An SWR meter indicates a perfect impedance match between the antenna and the feed line with a reading of **1 to 1**. (Hint* you want to match impedance, so 1 is exactly equal to 1.)

If you saw an SWR reading of 4:1, that would indicate an **impedance mismatch.**

You also have to consider power loss in the feed line that goes from your radio transceiver to your antenna. Because of current passing through the feed line, it loses power that **is converted to heat.**

Most solid-state amateur radio transmitters reduce output power as SWR increases **to protect the output amplifier transistors.**

There is an alternative way to measure impedance matching instead of SWR! A **directional wattmeter** could also be used instead of SWR to determine if a feed line and antenna are properly matched.

Common Problems

Don't talk too close to the microphone!

If you are told your FM handheld or mobile transceiver is over-deviating, **talk farther away from the microphone!**

Interference can be caused in many ways:

- **Fundamental overload**
- **Harmonics**
- **Spurious emissions**

If **the receiver is unable to reject strong signals outside the AM or FM band** it would cause a broadcast AM or FM radio to receive an amateur radio transmission unintentionally.

Radio transmissions could affect telephone or television!

You can eliminate interference from an amateur transmitter by putting an **RF filter on the telephone.**

An overload of a non-amateur radio or TV receiver by an amateur signal can be eliminated by **blocking the amateur signal with a filter at the antenna input of the affected receiver.**

If your neighbor says that your station's transmissions are interfering with their radio or TV reception, **make sure that your station is functioning properly and that it does not cause interference to your own radio or television when it is tuned to the same channel.**

The first step to avoid television interference is to **be sure all TV coaxial connectors are installed properly.**

If something in a neighbor's home is causing harmful interference to your station, there are several things to do:
- **Work with your neighbor to identify the offending device**
- **Politely inform your neighbor about the rules that prohibit the use of devices that cause interference**

- **Check your station and make sure it meets the standards of good amateur practice**

A **band-reject filter** can reduce overload to a VHF transceiver from a nearby FM broadcast station.

There could be several problems if you receive a report that your audio signal through the repeater is distorted or unintelligible:
- **Your transmitter is slightly off frequency**
- **Your batteries are running low**
- **You are in a bad location**

RF feedback can cause **reports of garbled, distorted, or unintelligible voice transmissions.**

Moisture contamination is the most common cause of failure of coaxial cables.

Ultraviolet light can damage the jacket and allow water to enter the cable.

The disadvantage of air core coaxial cable when compared to foam or solid dielectric types is **it requires special techniques to prevent water absorption.**

Terms

- **Sensitivity** - the ability of a receiver to detect the presence of a signal

- **Selectivity** - the ability of a receiver to discriminate between multiple signals
- **Modulation** - the combining speech with an RF carrier signal
- A Part 15 Device is **an unlicensed device that may emit low-powered radio signals on frequencies used by a licensed service.**

Chapter 7. Station Equipment

1) Which term describes the ability of a receiver to detect the presence of a signal?

A. Linearity
B. Sensitivity
C. Selectivity
D. Total Harmonic Distortion

2) What is a transceiver?

A. A type of antenna switch
B. A unit combining the functions of a transmitter and a receiver
C. A component in a repeater that filters out unwanted interference
D. A type of antenna matching network

3) Which of the following is used to convert a radio signal from one frequency to another?

A. Phase splitter
B. Mixer
C. Inverter
D. Amplifier

4) What can you do if you are told your FM handheld or mobile transceiver is over-deviating?

A. Talk louder into the microphone
B. Let the transceiver cool off
C. Change to a higher power level
D. Talk farther away from the microphone

5) What would cause a broadcast AM or FM radio to receive an amateur radio transmission unintentionally?

A. The receiver is unable to reject strong signals outside the AM or FM band
B. The microphone gain of the transmitter is turned up too high
C. The audio amplifier of the transmitter is overloaded
D. The deviation of an FM transmitter is set too low

6) Which of the following can cause radio frequency interference?

A. Fundamental overload
B. Harmonics
C. Spurious emissions
D. All of these choices are correct

7) What is the primary purpose of a dummy load?

A. To prevent transmitting signals over the air when making tests
B. To prevent over-modulation of a transmitter
C. To improve the efficiency of an antenna
D. To improve the signal-to-noise ratio of a receiver

8) Which of the following instruments can be used to determine if an antenna is resonant at the desired operating frequency?

A. A VTVM
B. An antenna analyzer
C. A Q meter
D. A frequency counter

9) Which instrument would you use to measure electric potential or electromotive force?

A. An ammeter
B. A voltmeter
C. A wavemeter
D. An ohmmeter

10) What is the correct way to connect a voltmeter to a circuit?

A. In series with the circuit
B. In parallel with the circuit
C. In quadrature with the circuit
D. In phase with the circuit

CHAPTER 8: MODULATION MODES

In ham radio, a mode is a way of using your ham radio. Some modes are used for talking, others for digital, and some for both. Let's talk about some common ham radio modes.

FM - Frequency Modulation

Everyone knows about Frequency Modulation (FM) mode because radio stations use it to transmit to our cars. In ham radio we can also use FM as a mode of communication. FM is a common voice mode (sound and talking). That's why **FM** modulation is most commonly used for VHF and UHF voice repeaters.

FM is also used for digital communications. **FM** modulation is most commonly used for VHF packet radio transmissions. We'll talk more about packet radio in digital communications.

SSB - Single Sideband

You had probably already heard of FM, but until ham radio you had probably never heard of SSB. SSB is another voice mode. Just like FM, it is used for sound and talking. **Single sideband** is a form of amplitude modulation instead of frequency.

An advantage of single sideband (SSB) over FM for voice transmissions is that **SSB signals have a narrow bandwidth**. For this reasons, many ham radio operators prefer using SSB for voice instead of FM. However, SSB can actually be finicky to tune because of the smaller bandwidth.

Single sideband (**SSB**) is used for long-distance (weak signal) contacts on the VHF and UHF bands.

The **upper sideband** is normally used for 10 meter HF, VHF, and UHF single-sideband communications.

FM	SSB
• Easier to tune • Larger bandwidth • Most commonly used for VHF and UHF voice repeaters • Used for VHF Packet radio transmissions	• Trickier to tune • Smaller bandwidth • Used for long-distance contacts on VHF and UHF bands • Upper side band is used for 10 meter HF, VHF, and UHF single-sideband communications

CW - Morse Code

You don't have to know Morse code for the exam anymore, but you should know some basics about what it is. Morse code, aka CW, is extremely simple - it just uses one on/off tone that is either long or short. Because it is so simple, **CW** has the narrowest bandwidth of the common modes. **150 Hz** is the appropriate maximum bandwidth required to transmit a CW signal.

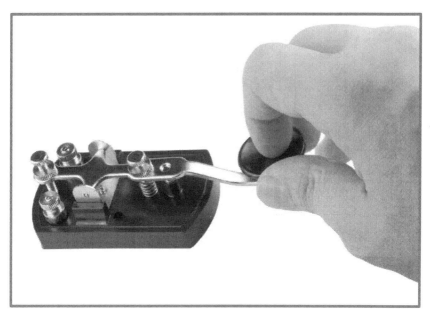

CW (aka Morse code) has the narrowest bandwidth of all common emissions at just 150Hz

An electronic keyer is **a device that assists in the manual sending of Morse code**. The picture here shows one in use, but you can also send morse code on almost all radios including handhelds.

International morse is used when sending CW. There are also more complex methods of digital communication.

Bandwidth Comparison of Common Modes

Using a different modes requires different amount of bandwidth. In order of least to greatest required bandwidth, here are some common modes ranked:

Lowest Bandwidth	1	Morse Code (CW)	~150Hz
	2	SSB	3kHz
	3	VHF Repeater FM Phone Signal	10 – 15 kHz
Highest Bandwidth	4	Analog TV Transmissions on 70cm band	About 6Mhz

Digital and Non-Voice Communication

Technology now allows us to send digital signals with ham radio. Digital modes use protocols to enable sending emails, text messages, images, and more. All of the following are digital communications modes:

- **Packet radio**
- **IEEE 802.11**
- **JT65**

Packet Radio

Packet radio is a cool way to send little "packets" of data in ham radio.

All of the following may be included in packet transmissions:

- **A check sum that permits error detection**

- **A header that contains the call sign of the station to which the information is being sent**
- **Automatic repeat request in case of error**

Technology has helped make packet radio better and more automated. "ARPS" is the **Automatic Packet Reporting System.**

APRS is used **to provide real-time tactical digital communications in conjunction with a map showing the locations of stations.**

Automatic Repeat Query (ARQ) technology helps make sure there are no errors in the transmission. The ARQ transmission system is **a digital scheme whereby the receiving station detects errors and sends a request to the sending station to retransmit the information.**

Voice Over Internet Protocol (VoIP)

Voice Over Internet Protocol (VoIP) is a **method of delivering voice communications over the internet using digital signals.**

Deliver voice signals over the internet

The Internet Radio Linking Project (IRLP) is a **technique to connect amateur radio systems, such as repeaters, via the internet using Voice Over Internet Protocol (VoIP).**

You can obtain a list of active nodes that use VoIP by:
- **By subscribing to an online service**
- **From online repeater lists maintained by the local repeater frequency coordinator**
- **From a repeater directory**

Access to some IRLP nodes is accomplished by using DTMF signals.

A gateway is an amateur radio station that is used to connect other amateur stations to the internet.

FT8

Ham radio technology is always improving. There is a new mode called FT8 that is making waves in the ham radio community because of it's speed and use in DXing. FT8 is **a digital mode capable of operating in low signal-to-noise conditions that transmit on 15-second intervals.**

PSK31

PSK31 is commonly used by ham radio operators to communicate in a keyboard-to-keyboard chat. The underlying modulation technology is called PSK, which stands for **Phase Shift Keying.**

Ham Radio Software

EchoLink systems also allow amateurs to communicate via the internet. Before you may use the EchoLink system to communicate using a repeater, **you must register your call sign and provide proof of license.**

Broadband-Hamnet ™ is **an amateur-radio-based data network using commercial Wi-Fi gear with modified firmware.**

GPS

Did you know that the Global Position System (GPS) works using radio waves?

A **Global Position System receiver** is used to provide data to the transmitter when sending automatic position reports from a mobile amateur radio station.

A grid locator is **a letter-number designator assigned to a geographic location**.

Amateur Television

Ham radio can be used to transmit television signals over amateur frequencies. One such television signal is NTCS. NTCS is **an analog fast scan color TV signal**.

DMR - Digital Mobile Radio

DMR has become popular in ham radio because of it's ease of use for digital communications. DMR (Digital Mobile Radio) is **a technique for time-multiplexing two digital voice signals on a single 12.5 kHz repeater channel**. This time multiplex technology provides a doubling of capacity.

Satellite & Space Communication

You can communicate with satellites in all 3 of the most common modes:
- **SSB**
- **FM**
- **CW/data**

The simplest way to receive information from a satellite is by listening to a satellite beacon. A satellite beacon is **a transmission from a satellite that contains status information.**

The beacons generally transmit **the health and status of the satellite.**

You could receive information about measurements on the satellite using telemetry (Hint* think measure to remember telemetry). **Anyone who can receive the telemetry signal** can receive telemetry from a space station.

All of the following are provided by satellite tracking programs:
- **Maps showing the real-time position of the satellite track over the earth**
- **The time, azimuth, and elevation of the start, maximum altitude, and end of a pass**
- **The apparent frequency of the satellite transmission, including effects of Doppler shift**

The impact of using too much effective radiated power on a satellite uplink is **blocking access by other users.**

Your signal strength on the downlink should be about the same as the beacon.

If a satellite is operating in mode U/V, **the satellite uplink is in the 70-centimeter band and the downlink is in the 2-meter band.**

Rotation of the satellite and its antennas causes spin fading of satellite signals.

The initials LEO tells you **the satellite is in a Low Earth Orbit.**

The Keplerian elements are inputs to a satellite tracking program.

Doppler shift is **an observed change in signal frequency caused by relative motion between the satellite and the earth station.**

WSJT digital software can be used for satellite/space communication. All of the following activities are supported by digital mode software in the WSJT suite:
- **Moonbounce or Earth-Moon-Earth**
- **Weak-signal propagation beacons**
- **Meteor scatter**

Ham Radio Contests

Contesting is trying to contact as many other stations as possible during a specified period of time.

When contacting another station in a radio contest, a good procedure is to **send only the minimum information needed for proper identification and the contest exchange.**

A fox hunt is another competition where you try to find a hidden transmitter. **Radio direction finding** is used to locate sources of noise interference or jamming. **A directional antenna** would be useful for a hidden transmitter hunt.

Chapter 8 Quiz

1) Which of the following is a form of amplitude modulation?

A. Spread spectrum
B. Packet radio
C. Single sideband
D. Phase shift keying (PSK)

2) What type of modulation is most commonly used for VHF packet radio transmissions?

A. FM
B. SSB
C. AM
D. PSK

3) Which type of voice mode is most often used for long-distance (weak signal) contacts on the VHF and UHF bands?

A. FM
B. DRM
C. SSB
D. PM

4) Which type of modulation is most commonly used for VHF and UHF voice repeaters?

A. AM
B. SSB
C. PSK
D. FM

5) Which of the following types of emission has the narrowest bandwidth?

A. FM voice
B. SSB voice
C. CW
D. Slow-scan TV

6) What telemetry information is typically transmitted by satellite beacons?

A. The signal strength of received signals
B. Time of day accurate to plus or minus 1/10 second
C. Health and status of the satellite
D. All of these choices are correct

7) What is the impact of using too much effective radiated power on a satellite uplink?

A. Possibility of commanding the satellite to an improper mode
B. Blocking access by other users
C. Overloading the satellite batteries
D. Possibility of rebooting the satellite control computer

8) Which of the following are provided by satellite tracking programs?

A. Maps showing the real-time position of the satellite track over the earth
B. The time, azimuth, and elevation of the start, maximum altitude, and end of a pass
C. The apparent frequency of the satellite transmission, including effects of Doppler shift
D. All of these choices are correct

9) What mode of transmission is commonly used by amateur radio satellites?

A. SSB
B. FM
C. CW/data
D. All of these choices are correct

10) What is a satellite beacon?

A. The primary transmit antenna on the satellite
B. An indicator light that shows where to point your antenna
C. A reflective surface on the satellite
D. A transmission from a satellite that contains status information

CHAPTER 9. ANTENNAS AND FEEDLINES

Antennas

Finally, it's time to discuss the actual hardware you need. You may be looking at a small radio with a small antenna (rubber-duck) or building a bigger long range antenna.

"Rubber-duck" antennas are not as effective as full antennas!

The little rubber duck antenna has a major disadvantage: **it does not transmit or receive as effectively as a full antenna.**

A beam antenna is an antenna that **concentrates signals in one direction.**

The Quad, Yagi, and dish antennas are all directional antennas.

A horizontally polarized antenna is an example of a simple dipole oriented parallel to the Earth's surface.

Inserting an inductor in the radiating portion of the antenna to make it electrically longer is a type of antenna loading.

To change a dipole antenna to make it resonant on a higher frequency, you can **shorten it.**

A half-wave dipole antenna radiates the strongest signal **broadside to the antenna.**

The gain of **the antenna is the increase in signal strength in a specified direction compared to a reference antenna. (Hint* gain means getting bigger)**

The approximate length in inches of a quarter-wavelength vertical antenna for 146 MHz is **19.** If you don't remember 19, you can calculate by determining that 146 MHz is located within the 2-meter band and that 2 meters is around 80 inches.

1/4 × 80 in=20 inches...**19 is the closest answer on the test.**

The approximate length, in inches, of a half-wavelength 6-meter dipole antenna, is **112.**

The advantage of using a properly mounted 5/8 wavelength antenna for VHF or UHF mobile service is **it has a lower radiation angle and more gain than a 1/4 wavelength antenna.** (Hint: It has *lower radiation angle and more gain!*)

5/8 wavelength antenna for VHF, UHF mobile service. Lower radiation angle, more gain!

Cables

It may seem simple, but choosing the correct cables is extremely important for a well operating station.

Coaxial cable the most common feed line selected for amateur radio antenna systems because **it is easy to use and requires few special installation considerations.**

The impedance of most coaxial cables used in amateur radio is **50 ohms.**

As the frequency of a signal passing through coaxial cable is increased **the loss increases.**

PL-259 type coax connectors **are commonly used at HF frequencies.**

(Hint* PL-259 coax connectors are not water tight!)

A Type N connector is most suitable for frequencies above 400 MHz.

Coax connectors exposed to weather should be sealed against water intrusion **to prevent an increase in feedline loss.**

The electrical difference between RG-58 and RG-8 coaxial cable is that **RG-8 cable has less loss at a given frequency.**

Air-insulated hard line has the lowest loss at VHF and UHF.

Tuning

Once you've got all your gear, it's time to make sure its tuned up.

The major function of an antenna tuner (antenna coupler) is **to match the antenna system impedance to the transceiver's output impedance.**

It is important to have low SWR when using coaxial cable feedline **to reduce signal loss.**

A **loose connection in an antenna or a feed line** can cause erratic changes in SWR readings.

The major function of an antenna tuner (antenna coupler) is **to match the antenna system impedance to the transceiver's output impedance.**

In Your Vehicle

A disadvantage of using a handheld VHF transceiver, with its integral antenna inside a vehicle is that **signals might not propagate well due to the shielding effect of the vehicle.**

Chapter 9 Quiz

1) What is a beam antenna?

A. An antenna built from aluminum I-beams
B. An omnidirectional antenna invented by Clarence Beam
C. An antenna that concentrates signals in one direction
D. An antenna that reverses the phase of received signals

2) Which of the following describes a type of antenna loading?

A. Inserting an inductor in the radiating portion of the antenna to make it electrically longer
B. Inserting a resistor in the radiating portion of the antenna to make it resonant
C. Installing a spring in the base of a mobile vertical antenna to make it more flexible
D. Strengthening the radiating elements of a beam antenna to better resist wind damage

3) Which of the following describes a simple dipole oriented parallel to the Earth's surface?

A. A ground-wave antenna
B. A horizontally polarized antenna
C. A rhombic antenna
D. A vertically polarized antenna

4) Why is it important to have low SWR when using coaxial cable feed line?

A. To reduce television interference
B. To reduce signal loss
C. To prolong antenna life
D. All of these choices are correct

5) What is the impedance of most coaxial cables used in amateur radio installations?

A. 8 ohms
B. 50 ohms
C. 600 ohms
D. 12 ohms

6) Why is coaxial cable the most common feed line selected for amateur radio antenna systems?

A. It is easy to use and requires few special installation considerations
B. It has less loss than any other type of feed line
C. It can handle more power than any other type of feed line
D. It is less expensive than any other type of feed line

7) Why should coax connectors exposed to the weather be sealed against water intrusion?

A. To prevent an increase in feed line loss
B. To prevent interference to telephones
C. To keep the jacket from becoming loose
D. All of these choices are correct

8) What can cause erratic changes in SWR readings?

A. The transmitter is being modulated
B. A loose connection in an antenna or a feed line
C. The transmitter is being over-modulated
D. Interference from other stations is distorting your signal

9) What is the electrical difference between RG-58 and RG-8 coaxial cable?

A. There is no significant difference between the two types
B. RG-58 cable has two shields
C. RG-8 cable has less loss at a given frequency
D. RG-58 cable can handle higher power levels

10) Which of the following types of feed line has the lowest loss at VHF and UHF?

A. 50-ohm flexible coax
B. Multi-conductor unbalanced cable
C. Air-insulated hard line
D. 75-ohm flexible coax

CHAPTER 10. ELECTRICAL SAFETY

Electrical Safety

Receiving an electrical current through your body presents many health hazards:
- **It may cause injury by heating tissue**
- **It may disrupt the electrical functions of cells**
- **It may cause involuntary muscle contractions**

There are many ways to guard against electrical shock at your station:
- **Use three-wire cords and plugs for all AC powered equipment**
- **Connect all AC powered station equipment to a common safety ground**
- **Use a circuit protected by a ground-fault interrupter**

In the United States, the green wire in a three-wire electrical plug is the **equipment ground**. Grounding prevents you from having a live electrical surface that could shock you severely!

For all external ground rods or earth connections, you should **bond them together with heavy wire or conductive strap.**

When installing grounding conductors used for lightning protection **sharp bends must be avoided.**

Excessive current can be dangerous, so you can use a fuse **to interrupt power in case of overload.**

Home-built equipment that is powered from 120V AC power circuits should always have **a fuse or circuit breaker in series with the AC hot conductor.**

It unwise to install a 20-ampere fuse in the place of a 5-ampere fuse because the **excessive current could cause a fire.**

Choose your fuse wisely or you could have an electrical fire on your hands!

You must take precautions with batteries as well. A safety hazard of a 12-volt storage battery is that **shorting the terminals can cause burns, fire, or an explosion.** If a lead-acid storage battery is charged or discharged too quickly, **the battery could overheat, give off flammable gas, or explode.**

Even when a power supply is turned off and disconnected, **you might receive an electric shock from the charge stored in large capacitors.**

An antenna should not be attached to a utility pole because **the antenna could contact high-voltage power lines.**

Radiation

Antennas create radiation that could potentially be dangerous.

If a person accidentally touched your antenna while you were transmitting **they might receive a painful RF burn.**

However, RF radiation is generally pretty safe compared to other more dangerous types of radiation. VHF and UHF radio signals are **non-ionizing radiation,** which doesn't do genetic damage like ionizing radiation. **RF radiation does not have sufficient energy to cause genetic damage.** But you could still get a nasty RF burn!

In order to make sure your tower is safe, you must limit radiation exposure to persons near the tower.

All of the following factors affect the RF exposure of people near an amateur antenna:
- **Frequency and power level of the RF field**
- **Distance from the antenna to a person**
- **Radiation pattern of the antenna**

Because **the human body absorbs more RF energy at some frequencies than at others,** exposure limits vary with frequency.

The maximum power level that an amateur radio station may use at VHF frequencies before an RF exposure evaluation is required is **50 watts PEP at the antenna.** (Hint: A lot of students get this question confused with the question asking what the peak envelope power output for

Technician class operators using frequencies above 30 MHz is. **It is NOT 1500!**)

50 MHz has the lowest value for Maximum Permissible Exposure Limit. (Hint: **50** is your magic number for questions regarding exposure limits)

To prevent exposure to RF radiation in excess of FCC-supplied limits you could **relocate your antenna.**

They are several ways to make sure your antenna is in compliance with the FCC:
- **By calculation based on FCC OET Bulletin 65**
- **By calculation based on computer modeling**
- **By measurement of field strength using calibrated equipment**

You can always make sure you are in FCC compliance by **re-evaluating the station whenever an item of equipment is changed.**

The duty cycle must be used in your calculations because it **affects the average exposure of people to radiation.** The duty cycle is defined as the percentage of time that a transmitter is transmitting. (Hint: The keyword here is **AVERAGE**. Don't be confused by the answer here that contains peak exposure.)

Obviously, if you're transmitting 100% of the time, you're going to get twice as much radiation as if you're transmitting only 50% of the time!

If the averaging time for exposure is 6 minutes, how much power density is permitted if the signal is present for 3 minutes and absent for 3 minutes rather than being present for the entire 6 minutes? **2 times as much.**

Towers

Members of a tower work team should wear a hard hat and safety glass **at all times when any work is being done on the tower.**

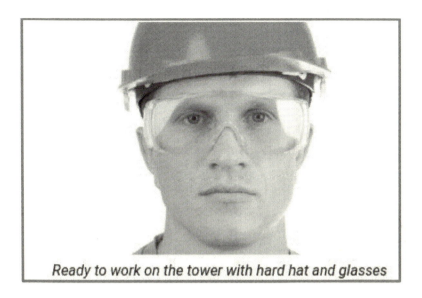
Ready to work on the tower with hard hat and glasses

A good precaution to observe before climbing an antenna tower is to **put on a carefully inspected climbing harness (fall arrester) and safety glasses.**

It is **NEVER** safe to climb a tower without a helper or observer.

Make sure to **look for and stay clear of any overhead electrical wires.**

Local electrical codes establish the proper grounding techniques for amateur towers.

Proper grounding is **separate eight-foot long ground rods for each tower leg, bonded to the tower and each other.**

When installing ground wires on a tower for electrical protection, **ensure that connections are short and direct.**

When installing devices for lightning protection in a coaxial cable feed line, **mount all of the protectors on a metal plate that is in turn connected to an external ground rod.**

Crank up towers are fast and convenient but you have to make sure to lock them. **They must not be climbed unless retracted or mechanical safety locking devices have been installed.**

A gin pole is used **to lift tower sections or antennas.**

A Few More Notes on Safety

When setting up an antenna, make sure you are far away enough from any power line **so that if the antenna falls unexpectedly, no part of it can come closer than 10 feet to the power wires.**

The purpose of safety wire through a turnbuckle used to tension guy lines is **to prevent loosening of the guy line from vibration.**

CONGRATULATIONS ON FINISHING YOUR STUDIES!

Once you are consistently passing with a score of 80% or better on your final exams (online and in the book) go ahead and register for your in-person exam. If you need helping finding an exam near you check out:

https://hamradioprep.com/find-a-ham-radio-license-exam/

That page will give you all of the latest information on exams in your area.

Don't forget to send a picture of you passing your exam to contact@hamradioprep.com and we'll send you a little gift on us :)

FINAL EXAM 1

1) Which of the following is a purpose of the Amateur Radio Service as stated in the FCC rules and regulations?

A. Providing personal radio communications for as many citizens as possible
B. Providing communications for international non-profit organizations
C. Advancing skills in the technical and communication phases of the radio art
D. All of these choices are correct

2) Which agency regulates and enforces the rules for the Amateur Radio Service in the United States?

A. FEMA
B. Homeland Security
C. The FCC
D. All of these choices are correct

3) What is the International Telecommunications Union (ITU)?

A. An agency of the United States Department of Telecommunications Management
B. A United Nations agency for information and communication technology issues
C. An independent frequency coordination agency
D. A department of the FCC

4) Which amateur radio stations may make contact with an amateur radio station on the International Space Station (ISS) using 2 meter and 70 cm band frequencies?

A. Only members of amateur radio clubs at NASA facilities
B. Any amateur holding a Technician or higher-class license
C. Only the astronaut's family members who are hams
D. Contacts with the ISS are not permitted on amateur radio frequencies

5) For which license classes are new licenses currently available from the FCC?

A. Novice, Technician, General, Advanced
B. Technician, Technician Plus, General, Advanced
C. Novice, Technician Plus, General, Advanced
D. Technician, General, Amateur Extra

6) Who may select a desired call sign under the vanity call sign rules?

A. Only a licensed amateur with a General or Amateur Extra class license
B. Only a licensed amateur with an Amateur Extra class license
C. Only a licensed amateur who has been licensed continuously for more than 10 years
D. Any licensed amateur

7) Which of the following is a common repeater frequency offset in the 2 meter band?

A. Plus or minus 5 MHz
B. Plus or minus 600 kHz
C. Plus or minus 500 kHz
D. Plus or minus 1 MHz

8) What is the national calling frequency for FM simplex operations in the 2 meter band?

A. 146.520 MHz
B. 145.000 MHz
C. 432.100 MHz
D. 446.000 MHz

9) What is the most common use of the "reverse split" function of a VHF/UHF transceiver?

A. Reduce power output
B. Increase power output
C. Listen on a repeater's input frequency
D. Listen on a repeater's output frequency

10) What should you do if another operator reports that your station's 2 meter signals were strong just a moment ago, but now they are weak or distorted?

A. Change the batteries in your radio to a different type
B. Turn on the CTCSS tone
C. Ask the other operator to adjust his squelch control
D. Try moving a few feet or changing the direction of your antenna if possible, as reflections may be causing multi-path distortion

11) Why might the range of VHF and UHF signals be greater in the winter?

A. Less ionospheric absorption
B. Less absorption by vegetation
C. Less solar activity
D. Less tropospheric absorption

12) What is the name for the distance a radio wave travels during one complete cycle?

A. Wave speed
B. Waveform
C. Wavelength
D. Wave spread

13) What must be considered to determine the minimum current capacity needed for a transceiver power supply?

A. Efficiency of the transmitter at full power output
B. Receiver and control circuit power
C. Power supply regulation and heat dissipation
D. All of these choices are correct

14) How might a computer be used as part of an amateur radio station?

A. For logging contacts and contact information
B. For sending and/or receiving CW
C. For generating and decoding digital signals
D. All of these choices are correct

15) Electrical current is measured in which of the following units?

A. Volts
B. Watts
C. Ohms
D. Amperes

16) Electrical power is measured in which of the following units?

A. Volts
B. Watts
C. Ohms
D. Amperes

17) How much voltage does a mobile transceiver typically require?

A. About 12 volts
B. About 30 volts
C. About 120 volts
D. About 240 volts

18) Which of the following is a good electrical conductor?

A. Glass
B. Wood
C. Copper
D. Rubber

19) What electrical component opposes the flow of current in a DC circuit?

A. Inductor
B. Resistor
C. Voltmeter
D. Transformer

20) What type of component is often used as an adjustable volume control?

A. Fixed resistor
B. Power resistor
C. Potentiometer
D. Transformer

21) What class of electronic components uses a voltage or current signal to control current flow?

A. Capacitors
B. Inductors
C. Resistors
D. Transistors

22) What electronic component allows current to flow in only one direction?

A. Resistor
B. Fuse
C. Diode
D. Driven element

23) Which term describes the ability of a receiver to detect the presence of a signal?

A. Linearity
B. Sensitivity
C. Selectivity
D. Total Harmonic Distortion

24) What is a transceiver?

A. A type of antenna switch
B. A unit combining the functions of a transmitter and a receiver
C. A component in a repeater that filters out unwanted interference
D. A type of antenna matching network

25) How can overload of a non-amateur radio or TV receiver by an amateur signal be reduced or eliminated?

A. Block the amateur signal with a filter at the antenna input of the affected receiver
B. Block the interfering signal with a filter on the amateur transmitter
C. Switch the transmitter from FM to SSB
D. Switch the transmitter to a narrow-band mode

26) Which of the following actions should you take if a neighbor tells you that your station's transmissions are interfering with their radio or TV reception?

A. Make sure that your station is functioning properly and that it does not cause interference to your own radio or television when it is tuned to the same channel
B. Immediately turn off your transmitter and contact the nearest FCC office for assistance
C. Tell them that your license gives you the right to transmit and nothing can be done to reduce the interference
D. Install a harmonic doubler on the output of your transmitter and tune it until the interference is eliminated

27) Which of the following is a form of amplitude modulation?
A. Spread spectrum
B. Packet radio
C. Single sideband
D. Phase shift keying (PSK)

28) What type of modulation is most commonly used for VHF packet radio transmissions?

A. FM
B. SSB
C. AM
D. PSK

29) What mode of transmission is commonly used by amateur radio satellites?

A. SSB
B. FM
C. CW/data
D. All of these choices are correct

30) What is a satellite beacon?

A. The primary transmit antenna on the satellite
B. An indicator light that shows where to point your antenna
C. A reflective surface on the satellite
D. A transmission from a satellite that contains status information

31) Which of the following describes a type of antenna loading?

A. Inserting an inductor in the radiating portion of the antenna to make it electrically longer
B. Inserting a resistor in the radiating portion of the antenna to make it resonant
C. Installing a spring in the base of a mobile vertical antenna to make it more flexible
D. Strengthening the radiating elements of a beam antenna to better resist wind damage

32) Which of the following describes a simple dipole oriented parallel to the Earth's surface?

A. A ground-wave antenna
B. A horizontally polarized antenna
C. A rhombic antenna
D. A vertically polarized antenna

33) Which of the following is a safety hazard of a 12-volt storage battery?

A. Touching both terminals with the hands can cause electrical shock
B. Shorting the terminals can cause burns, fire, or an explosion
C. RF emissions from the battery
D. All of these choices are correct

34) What health hazard is presented by electrical current flowing through the body?

A. It may cause injury by heating tissue
B. It may disrupt the electrical functions of cells
C. It may cause involuntary muscle contractions
D. All of these choices are correct

35) In the United States, what is connected to the green wire in a three-wire electrical AC plug?

A. Neutral
B. Hot
C. Equipment ground
D. The white wire

FINAL EXAM 2

1) What should be done to all external ground rods or earth connections?

A. Waterproof them with silicone caulk or electrical tape
B. Keep them as far apart as possible
C. Bond them together with heavy wire or conductive strap
D. Tune them for resonance on the lowest frequency of operation

2) What can happen if a lead-acid storage battery is charged or discharged too quickly?

A. The battery could overheat, give off flammable gas, or explode
B. The voltage can become reversed
C. The memory effect will reduce the capacity of the battery
D. All of these choices are correct

3) What kind of hazard might exist in a power supply when it is turned off and disconnected?

A. Static electricity could damage the grounding system
B. Circulating currents inside the transformer might cause damage
C. The fuse might blow if you remove the cover
D. You might receive an electric shock from the charge stored in large capacitors

4) Why is coaxial cable the most common feed line selected for amateur radio antenna systems?

A. It is easy to use and requires few special installation considerations
B. It has less loss than any other type of feed line
C. It can handle more power than any other type of feed line
D. It is less expensive than any other type of feed line

5) What is the major function of an antenna tuner (antenna coupler)?

A. It matches the antenna system impedance to the transceiver's output impedance
B. It helps a receiver automatically tune in weak stations
C. It allows an antenna to be used on both transmit and receive
D. It automatically selects the proper antenna for the frequency band being used

6) What is an ARQ transmission system?

A. A special transmission format limited to video signals
B. A system used to encrypt command signals to an amateur radio satellite
C. A digital scheme whereby the receiving station detects errors and sends a request to the sending station to retransmit the information
D. A method of compressing the data in a message so more information can be sent in a shorter time

7) Which of the following best describes Broadband-Hamnet(TM), also referred to as a high-speed multi-media network?

A. An amateur-radio-based data network using commercial Wi-Fi gear with modified firmware
B. A wide-bandwidth digital voice mode employing DRM protocols
C. A satellite communications network using modified commercial satellite TV hardware
D. An internet linking protocol used to network repeaters

8) What is FT8?

A. A wideband FM voice mode
B. A digital mode capable of operating in low signal-to-noise conditions that transmits on 15-second intervals
C. An eight channel multiplex mode for FM repeaters
D. A digital slow scan TV mode with forward error correction and automatic color compensation

9) What is an electronic keyer?

A. A device for switching antennas from transmit to receive
B. A device for voice activated switching from receive to transmit
C. A device that assists in manual sending of Morse code
D. An interlock to prevent unauthorized use of a radio

10) What is the characteristic appearance of a cold solder joint?

A. Dark black spots
B. A bright or shiny surface
C. A grainy or dull surface
D. A greenish tint

11) What is probably happening when an ohmmeter, connected across an unpowered circuit, initially indicates a low resistance and then shows increasing resistance with time?

A. The ohmmeter is defective
B. The circuit contains a large capacitor
C. The circuit contains a large inductor
D. The circuit is a relaxation oscillator

12) Which of the following precautions should be taken when measuring circuit resistance with an ohmmeter?

A. Ensure that the applied voltages are correct
B. Ensure that the circuit is not powered
C. Ensure that the circuit is grounded
D. Ensure that the circuit is operating at the correct frequency

13) Which of the following precautions should be taken when measuring high voltages with a voltmeter?

A. Ensure that the voltmeter has very low impedance
B. Ensure that the voltmeter and leads are rated for use at the voltages to be measured
C. Ensure that the circuit is grounded through the voltmeter
D. Ensure that the voltmeter is set to the correct frequency

14) Which of the following is combined with an inductor to make a tuned circuit?

A. Resistor
B. Zener diode
C. Potentiometer
D. Capacitor

15) What is the name of a device that combines several semiconductors and other components into one package?

A. Transducer
B. Multi-pole relay
C. Integrated circuit
D. Transformer

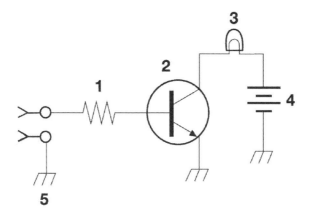

Figure T-1

16) What is the function of component 2 in Figure T1?

A. Give off light when current flows through it
B. Supply electrical energy
C. Control the flow of current
D. Convert electrical energy into radio waves

17) Which of the following is a resonant or tuned circuit?

A. An inductor and a capacitor connected in series or parallel to form a filter
B. A type of voltage regulator
C. A resistor circuit used for reducing standing wave ratio
D. A circuit designed to provide high-fidelity audio

18) What happens to current at the junction of two components in series?

A. It divides equally between them
B. It is unchanged
C. It divides based on the on the value of the components
D. The current in the second component is zero

19) What happens to current at the junction of two components in parallel?

A. It divides between them dependent on the value of the components
B. It is the same in both components
C. Its value doubles
D. Its value is halved

20) What is the voltage across each of two components in series with a voltage source?

A. The same voltage as the source
B. Half the source voltage
C. It is determined by the type and value of the components
D. Twice the source voltage

21) What is the voltage across each of two components in parallel with a voltage source?

A. It is determined by the type and value of the components
B. Half the source voltage
C. Twice the source voltage
D. The same voltage as the source

22) What is the advantage of having multiple receive bandwidth choices on a multimode transceiver?

A. Permits monitoring several modes at once
B. Permits noise or interference reduction by selecting a bandwidth matching the mode
C. Increases the number of frequencies that can be stored in memory
D. Increases the amount of offset between receive and transmit frequencies

23) Which of the following is an appropriate receive filter bandwidth for minimizing noise and interference for SSB reception?

A. 500 Hz
B. 1000 Hz
C. 2400 Hz
D. 5000 Hz

24) What is generally the best time for long-distance 10 meter band propagation via the F layer?

A. From dawn to shortly after sunset during periods of high sunspot activity
B. From shortly after sunset to dawn during periods of high sunspot activity
C. From dawn to shortly after sunset during periods of low sunspot activity
D. From shortly after sunset to dawn during periods of low sunspot activity

25) Which of the following bands may provide long distance communications during the peak of the sunspot cycle?

A. 6 or 10 meter bands
B. 23 centimeter band
C. 70 centimeter or 1.25 meter bands
D. All of these choices are correct

26) Why do VHF and UHF radio signals usually travel somewhat farther than the visual line of sight distance between two stations?

A. Radio signals move somewhat faster than the speed of light
B. Radio waves are not blocked by dust particles
C. The Earth seems less curved to radio waves than to light
D. Radio waves are blocked by dust particles

27) What information is contained in the preamble of a formal traffic message?

A. The email address of the originating station
B. The address of the intended recipient
C. The telephone number of the addressee
D. The information needed to track the message

28) What is meant by the term "check," in reference to a formal traffic message?

A. The number of words or word equivalents in the text portion of the message
B. The value of a money order attached to the message
C. A list of stations that have relayed the message
D. A box on the message form that indicates that the message was received and/or relayed

29) What is the Amateur Radio Emergency Service (ARES)?

A. Licensed amateurs who have voluntarily registered their qualifications and equipment for communications duty in the public service
B. Licensed amateurs who are members of the military and who voluntarily agreed to provide message handling services in the case of an emergency
C. A training program that provides licensing courses for those interested in obtaining an amateur license to use during emergencies
D. A training program that certifies amateur operators for membership in the Radio Amateur Civil Emergency Service

30) What method of call sign identification is required for a station transmitting phone signals?

A. Send the call sign followed by the indicator RPT
B. Send the call sign using a CW or phone emission
C. Send the call sign followed by the indicator R
D. Send the call sign using only a phone emission

31) Which of the following formats of a self-assigned indicator is acceptable when identifying using a phone transmission?

A. KL7CC stroke W3
B. KL7CC slant W3
C. KL7CC slash W3
D. All of these choices are correct

32) Which of the following restrictions apply when a non-licensed person is allowed to speak to a foreign station using a station under the control of a Technician class control operator?

A. The person must be a U.S. citizen
B. The foreign station must be one with which the U.S. has a third-party agreement
C. The licensed control operator must do the station identification
D. All of these choices are correct

33) What is meant by the term Third Party Communications?

A. A message from a control operator to another amateur station control operator on behalf of another person
B. Amateur radio communications where three stations are in communications with one another
C. Operation when the transmitting equipment is licensed to a person other than the control operator
D. Temporary authorization for an unlicensed person to transmit on the amateur bands for technical experiments

34) What type of amateur station simultaneously retransmits the signal of another amateur station on a different channel or channels?

A. Beacon station
B. Earth station
C. Repeater station
D. Message forwarding station

35) Who is accountable should a repeater inadvertently retransmit communications that violate the FCC rules?

A. The control operator of the originating station
B. The control operator of the repeater
C. The owner of the repeater
D. Both the originating station and the repeater owner

FINAL EXAM 3

1) When is it permissible to transmit messages encoded to hide their meaning?

A. Only during contests
B. Only when operating mobile
C. Only when transmitting control commands to space stations or radio control craft
D. Only when frequencies above 1280 MHz are used

2) Under what conditions is an amateur station authorized to transmit music using a phone emission?

A. When incidental to an authorized retransmission of manned spacecraft communications
B. When the music produces no spurious emissions
C. When the purpose is to interfere with an illegal transmission
D. When the music is transmitted above 1280 MHz

3) When may amateur radio operators use their stations to notify other amateurs of the availability of equipment for sale or trade?

A. When the equipment is normally used in an amateur station and such activity is not conducted on a regular basis
B. When the asking price is $100.00 or less
C. When the asking price is less than its appraised value
D. When the equipment is not the personal property of either the station licensee or the control operator or their close relatives

4) What, if any, are the restrictions concerning transmission of language that may be considered indecent or obscene?

A. The FCC maintains a list of words that are not permitted to be used on amateur frequencies
B. Any such language is prohibited
C. The ITU maintains a list of words that are not permitted to be used on amateur frequencies
D. There is no such prohibition

5) What types of amateur stations can automatically retransmit the signals of other amateur stations?

A. Auxiliary, beacon, or Earth stations
B. Repeater, auxiliary, or space stations
C. Beacon, repeater, or space stations
D. Earth, repeater, or space stations

6) In which of the following circumstances may the control operator of an amateur station receive compensation for operating that station?

A. When the communication is related to the sale of amateur equipment by the control operator's employer
B. When the communication is incidental to classroom instruction at an educational institution
C. When the communication is made to obtain emergency information for a local broadcast station
D. All of these choices are correct

7) What is meant by "repeater offset?"

A. The difference between a repeater's transmit frequency and its receive frequency
B. The repeater has a time delay to prevent interference
C. The repeater station identification is done on a separate frequency
D. The number of simultaneous transmit frequencies used by a repeater

8) What is the meaning of the procedural signal "CQ"?

A. Call on the quarter hour
B. A new antenna is being tested (no station should answer)
C. Only the called station should transmit
D. Calling any station

9) What type of tones are used to control repeaters linked by the Internet Relay Linking Project (IRLP) protocol?

A. DTMF
B. CTCSS
C. EchoLink
D. Sub-audible

10) What term is commonly used to describe the rapid fluttering sound sometimes heard from mobile stations that are moving while transmitting?

A. Flip-flopping
B. Picket fencing
C. Frequency shifting
D. Pulsing

11) What type of wave carries radio signals between transmitting and receiving stations?

A. Electromagnetic
B. Electrostatic
C. Surface acoustic
D. Ferromagnetic

12) Which of the following is a likely cause of irregular fading of signals received by ionospheric reflection?

A. Frequency shift due to Faraday rotation
B. Interference from thunderstorms
C. Random combining of signals arriving via different paths
D. Intermodulation distortion

13) Which of the following could be used to remove power line noise or ignition noise?

A. Squelch
B. Noise blanker
C. Notch filter
D. All of these choices are correct

14) Which of the following is a use for the scanning function of an FM transceiver?

A. To check incoming signal deviation
B. To prevent interference to nearby repeaters
C. To scan through a range of frequencies to check for activity
D. To check for messages left on a digital bulletin board

15) What is another way to specify a radio signal frequency of 1,500,000 hertz?

A. 1500 kHz
B. 1500 MHz
C. 15 GHz
D. 150 kHz

16) How many volts are equal to one kilovolt?

A. One one-thousandth of a volt
B. One hundred volts
C. One thousand volts
D. One million volts

17) What is the ability to store energy in an electric field called?

A. Inductance
B. Resistance
C. Tolerance
D. Capacitance

18) What is the basic unit of capacitance?

A. The farad
B. The ohm
C. The volt
D. The henry

19) What electrical parameter is controlled by a potentiometer?

A. Inductance
B. Resistance
C. Capacitance
D. Field strength

20) What electrical component stores energy in an electric field?

A. Resistor
B. Capacitor
C. Inductor
D. Diode

21) What are the names of the two electrodes of a diode?

A. Plus and minus
B. Source and drain
C. Anode and cathode
D. Gate and base

22) Which of the following could be the primary gain-producing component in an RF power amplifier?

A. Transformer
B. Transistor
C. Reactor
D. Resistor

23) What is the name of a circuit that generates a signal at a specific frequency?

A. Reactance modulator
B. Product detector
C. Low-pass filter
D. Oscillator

24) What device converts the RF input and output of a transceiver to another band?

A. High-pass filter
B. Low-pass filter
C. Transverter
D. Phase converter

25) What might be a problem if you receive a report that your audio signal through the repeater is distorted or unintelligible?

A. Your transmitter is slightly off frequency
B. Your batteries are running low
C. You are in a bad location
D. All of these choices are correct

26) What is a symptom of RF feedback in a transmitter or transceiver?

A. Excessive SWR at the antenna connection
B. The transmitter will not stay on the desired frequency
C. Reports of garbled, distorted, or unintelligible voice transmissions
D. Frequent blowing of power supply fuses

27) Which of the following types of emission has the narrowest bandwidth?

A. FM voice
B. SSB voice
C. CW
D. Slow-scan TV

28) Which sideband is normally used for 10 meter HF, VHF, and UHF single-sideband communications?

A. Upper sideband
B. Lower sideband
C. Suppressed sideband
D. Inverted sideband

29) What is the impact of using too much effective radiated power on a satellite uplink?

A. Possibility of commanding the satellite to an improper mode
B. Blocking access by other users
C. Overloading the satellite batteries
D. Possibility of rebooting the satellite control computer

30) Which of the following are provided by satellite tracking programs?

A. Maps showing the real-time position of the satellite track over the earth
B. The time, azimuth, and elevation of the start, maximum altitude, and end of a pass
C. The apparent frequency of the satellite transmission, including effects of Doppler shift
D. All of these choices are correct

31) What is a disadvantage of using a handheld VHF transceiver, with its integral antenna, inside a vehicle?

A. Signals might not propagate well due to the shielding effect of the vehicle
B. It might cause the transceiver to overheat
C. The SWR might decrease, decreasing the signal strength
D. All of these choices are correct

32) Which of the following connectors is most suitable for frequencies above 400 MHz?

A. A UHF (PL-259/SO-239) connector
B. A Type N connector
C. An RS-213 connector
D. A DB-25 connector

33) Which of the following is an important safety precaution to observe when putting up an antenna tower?

A. Wear a ground strap connected to your wrist at all times
B. Insulate the base of the tower to avoid lightning strikes
C. Look for and stay clear of any overhead electrical wires
D. All of these choices are correct

34) What is the purpose of a gin pole?

A. To temporarily replace guy wires
B. To be used in place of a safety harness
C. To lift tower sections or antennas
D. To provide a temporary ground

35) Which of the following is true when installing grounding conductors used for lightning protection?

A. Only non-insulated wire must be used
B. Wires must be carefully routed with precise right-angle bends
C. Sharp bends must be avoided
D. Common grounds must be avoided

CHAPTER QUIZ ANSWERS

Chapter 1. Introduction to Amateur Radio

1) C 2) D 3) B 4) D 5) A 6) D 7) A 8) D 9) D 10) B

Chapter 2. Operating Procedures

1) A 2) C 3) A 4) D 5) D 6) A 7) A 8) C 9) A 10) D

Chapter 3. Radio Wave Characteristics

1) D 2) B 3) C 4) C 5) A 6) C 7) C 8) C 9) B 10) B

Chapter 4. Amateur Radio Practices

1) C 2) A 3) A 4) B 5) C 6) D 7) D 8) A 9) C 10) A

Chapter 5. Electrical Principles

1) C 2) A 3) C 4) D 5) A 6) D 7) D 8) A 9) A 10) C

Chapter 6. Electrical Components

1) A 2) B 3) D 4) B 5) B 6) A 7) B 8) A 9) D 10) B

Chapter 7. Station Equipment

1) B 2) B 3) B 4) D 5) A 6) D 7) A 8) B 9) B 10) B

Chapter 8. Modulation Modes

1) C 2) A 3) C 4) D 5) C 6) C 7) B 8) D 9) D 10) D

Chapter 9. Antennas and Feedlines

1) C 2) A 3) D 4) B 5) B 6) A 7) A 8) B 9) C 10) C

Chapter 10. Electrical Safety

1) B 2) D 3) C 4) C 5) C 6) D 7) C 8) C 9) D 10) C

ANSWER KEY FOR FINAL EXAMS

Final Exam 1

1) C 2) C 3) B 4) B 5) D 6) D 7) B 8) A 9) C 10) D 11) B 12) C 13) D 14) D
15) D 16) B 17) A 18) C 19) B 20) C 21) D 22) C 23) B 24) B 25) A 26) A
27) C 28) A 29) D 30) D 31) A 32) B
33) B 34) D 35) C

Final Exam 2

1) C 2) A 3) D 4) A 5) A 6) C 7) A 8) B 9) C 10) C 11) B 12) B 13) B 14) D
15) C 16) C 17) A 18) B 19) A 20) C 21) D 22) B 23) C 24) A 25) A 26) C
27) D 28) A 29) A 30) B 31) D 32) B 33) A 34) C 35) A

Final Exam 3

1) C 2) A 3) A 4) B 5) B 6) B 7) A 8) D 9) A 10) B 11) A 12) C 13) B 14) C 15) A 16) C 17) D 18) A 19) B 20) B 21) C 22) B 23) D 24) C 25) D 26) C 27) C 28) A 29) B 30) D 31) A 32) B 33) C 34) C 35) C

Your Coupon Code for Access to the Online Course is:

UAX071489

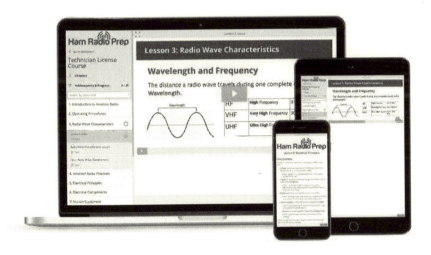

It Includes:

- Unlimited Practice Exams and Quizzes
- Videos to Help Your Learning
- Full Color Illustrations

Made in the USA
Middletown, DE
22 October 2020